# Writing Science through Critical Thinking

The Jones and Bartlett Series in
Logic, Critical Thinking, and Scientific Method

Gary Jason, Editor

Nancy Carrick and Lawrence Finsen, University of Redlands
*The Persuasive Pen: An Integrated Approach to Reasoning and Writing*

John Heil, Davidson College
*First-Order Logic: A Concise Introduction*

Gary Jason, San Diego State University
*Introduction to Logic*

Marilyn F. Moriarty, Hollins College
*Writing Science through Critical Thinking*

---

Other Titles of Interest

Liane Reif-Lehrer, Tech-Write Consultants/ERIMON Associates
*Grant Application Writers Handbooks, Third Edition*

Liane Reif-Lehrer, Tech-Write Consultants/ERIMON Associates
*Getting Funded: It Takes More than Just a Good Idea* (Videos)

John F. Petrik, Rockhurst College
*Reading and Writing Arguments: A Primer for Beginning Students*

Robin White, California State Polytechnic University
*The Winning Writer: Studies in the Art of Self-Expression*

Deborah St. James, Medical Editor, with Howard Spiro, Yale University
School of Medicine
*Writing and Speaking for Excellence: A Guide for Physicians*

# Writing Science through Critical Thinking

Marilyn F. Moriarty
*Hollins College*

**Jones and Bartlett Publishers**
*Sudbury, Massachusetts*
Boston      London      Singapore

*Editorial, Sales, and Customer Service Offices*
Jones and Bartlett Publishers
40 Tall Pine Drive
Sudbury, MA 01776
1-508-433-5000     info@jbpub.com
1-800-832-0034     http://www.jbpub.com

Jones and Bartlett Publishers International
Barb House, Barb Mews
London W6 7PA
UK

**Library of Congress Cataloging-in-Publication Data**

Moriarty, Marilyn F.
    Writing science through critical thinking / Marilyn F. Moriarty.
        p.   cm. — (The Jones and Bartlett series in logic, critical thinking, and scientific method)
    Includes bibliographical references and index.
    ISBN 0-86720-510-5
    1. Technical Writing.  2. Critical thinking.  I. Title.  II. Series.
T11.M573 1996
808.0666—dc20                                                                96-28447
                                                                                 CIP

**Credits**
*Acquisitions Editors:* Arthur C. Bartlett and Nancy E. Bartlett
*Manufacturing Manager:* Dana L. Cerrito
*Editorial Production Service:* Books in Philosophy
*Typesetting:* Books in Philosophy
*Cover Design:* Hannus Design Associates
*Printing and Binding:* Edwards Brothers
*Cover Printing:* John P. Pow Company, Inc.

Printed in the United States of America
99 98 97 96   10 9 8 7 6 5 4 3 2 1

Dedicated to Robert A. Varsano
(August 19, 1953–July 24, 1995)

# Contents

# Foreword

*Writing Science through Critical Thinking* has its roots in an academic alliance dedicated to the recruitment and retention of underrepresented students in the sciences. In California, this alliance was implemented in 1991 at the University of California, Irvine (UCI), through a cooperative agreement between the National Science Foundation and UCI, and identified as the California Alliance for Minority Participation (CAMP). CAMP evolved from an earlier initiative, the Howard Hughes Undergraduate Biological Sciences Minority Advanced Research and Training Program with additional support from the National Institutes of Health. Under the leadership of Professor Eloy Ródriguez, professor of developmental and cell biology, and principal investigator, and Professor Luis Villarreal, professor of molecular biology and biochemistry, and regional director, CAMP emphasized the importance of fostering scientific writing skills for science students as a key factor underlying their academic success. Ródriguez and Villarreal recognized that writing about science comprised a necessary complement to knowledge only when those results are published. In order to produce a cadre of publishing researchers, a writing component accompanied the hands-on research units.

In 1992, support from Vice Chancellor Manuel N. Gómez and UCI's science faculty brought forth the summer academy's writing component, which formed the genesis of this book. The first draft was enthusiastically received, and funding was provided for revision. It soon became apparent that the emphasis on writing skills pedagogically reinforced the interdisciplinary linkages between science curriculum, critical thinking, and scientific writing. Thus, promoting student writing skills served the dual function of preparing students for both their immediate future within the university and their professional development as potential scientists contributing to the body of scientific knowledge.

*Laurel L. Wilkening*
*Chancellor, University of California, Irvine, and*
*Principal Investigator, CAMP*

# Preface

Scientific writing, like so much in academia, is an acquired skill. It seems that transmitting writing skill is essential among the various aspects of scientific training. Most of my colleagues will agree that the one chore eliciting most consternation is the effort it takes to transform a young graduate student's first draft into a publishable manuscript. Dr. Marilyn Moriarty makes great strides in helping students acquire the keen thinking and writing competencies required for success in the science disciplines. For science writers, that means an objective tone built from words that are concrete, specific, and concise. Few students, if any, will succeed in science if they fail to establish themselves through writing. That is the ultimate premise we at the University of California, Irvine, cite in our science writing component: *establish yourself through writing.*

As I consider how I came to establish myself, to acquire some of this skill, I am struck by how long and disjointed a process it was. Today, as I observe the struggle among my own students, I am also struck by the universality of my experience. Like most undergraduate science students, I was introduced to the process of scientific writing through the established practice of writing lab reports, an affair in which one attempts to obtain the expected results and neatly recount the experiment. Such a rote practice, however, proved inadequate when attempting to construct an original idea.

Teaching has shown me that when beginning science students are faced with concepts or observations which have not been previously reported, many will be paralyzed by the task of creating an original argument. They will either parrot the ideas of others or spin their mental wheels. Additionally, from my long standing interest in minority and women science education, I know that these students often have greater difficulty with science writing, due, I believe, to their limited exposure to the process. During the development of various programs at Irvine, it became clear that the problems of linking writing and thinking were not limited to minority students but applicable to all science students, even most graduate students.

Before I had joined the faculty ranks, when I was asked to write my first original paper as a graduate student, my inadequacy in conveying scientific thought and analysis was made very apparent. My mentor asked that I work through the first draft unassisted. Although my results were clear and significant, and my objectives well focused, my ability to convey them effectively was almost absent, as my mentor's critique made painfully obvious. Among other problems, I had difficulty distinguishing between the results and the discussion sections, so they tended to be repetitious. Thus, I learned that writing involved an evolutionary process of attempts and responses to critical editing. Now we have this welcome student guide that directs students to set parameters, self-edit, and revise.

*Writing Science through Critical Thinking* takes the pain out of the process and expedites skill acquisition for today's competitive students. It short-circuits the tedium of false starts through clear direction and description of a scientific paper, from the abstract to the bibliography. From Chapter 1, *Thinking Science*, to the library search, Dr. Moriarty leads students out of the quagmire and into the clarity of patterns and process analysis.

Mostly, the problems lie not with grammar, syntax, style, or vocabulary, but rather with *science thinking* itself or, more specifically, the ordering of thought through writing. This is the challenge: to reconstruct, to disassemble into coherent parts, and to reorder. Students must work through the confusion themselves, and they often underestimate the amount of labor that is needed for scholarly writing. As with all scientific endeavors, one must organize a strategy that reflects focus and initiative. Hard to achieve. Most professors prefer to start over. But it appears that students must make the effort in order to "see" how their own thinking should be reordered.

A major problem in teaching *science writing* is that one must also learn *science thinking*, and most would strongly argue that only another scientist can do this. Whose responsibility is it to prepare the student for both aspects of their discipline? Is it the English Department's only? How many scientists have time to teach all their students to write? Yet, it is the singular crucial skill for their own successful careers.

I also believe that the underlying skill, what we call critical thinking, must be intrinsically engaged in the process—the reasoning ability to differentiate *what one sees* from *what one thinks one sees* from *what one thinks it means*. Dr. Moriarty's book leads students through the minefield of interpretation and objective analysis with a pragmatic approach.

Moriarty clearly defines the problems and applies her longstanding interest in the history of science, along with her extensive background in teaching technical writing. She has produced a book in which the link between seeing and describing is methodically presented. The exercises presented allow the students to "see" their own writing. The clarity of thought and the clarity of writing are intertwined. Students are encouraged to observe, describe, interpret. There is no other way. Our science majors at

UCI survived the pilot course intact and produced some of the finest abstracts ever produced by freshmen and transfer students.

At the University of California, Irvine, as the director, along with Dr. Eloy Ródriguez, and developer of various minority and women-student science programs, I began considering how students learn science. The key program most related to the development of science writing course was the California Alliance for Minority Participation (CAMP) funded through a cooperative agreement with the National Sciences Foundation. Begun in 1991 at UCI, this program links major research universities with state colleges and community colleges to address the problems of science education and retention for minority students. Moriarty was the first writing instructor for CAMP. At that time, we could find no textbook or reference book which approached the problems of science writing; Moriarty constructed her own handbook. CAMP, however, that provided the context and the catalyst for this book. In the following academic year, I adopted it for interim use in a formal course I offer in science writing, intended to fill the upper-division writing requirement. These were to be students who had done a considerable amount of research, many of whom were also participating in the honors program.

This class was taught in a computer lab with full library and internet access. Students performed a considerable amount of literature research using Medline and other library databases. They also had access to word processors. Their drafts were reviewed in stages and the final manuscript underwent a full review by several science faculty for possible publication in the *Journal of Undergraduate Research in the Biological Sciences* here at UCI. Thus, both writing and publishing in the sciences was incorporated.

Since the development of the first class, I have formed another in which students do not need prior research experience. Because it would be very difficult for most students at a large public college or university to do independent research, a writing class for these students should not be dependent on such experience. The computer network, including Genbank, was used to simulate original research. Thus, although this book is written for students with research experience, it works well for classes requiring library and/or simulated research. In either case, the students will be asking a catalog of questions: What is it made of? How does it work? What is the significance? Somewhere between objectivity and concrete language, they will write the answers.

*Luis P. Villarreal*
*Department of Molecular Biology and Biochemistry*
*University of California, Irvine*
*Irvine, CA 92717-3900*

signature of Dr. Hartz's perspicuity in the exercise on Galileo's "The Starry Messenger." Thanks enough are not enough thanks for her contribution.

Thanks also to the following for permission to use their unpublished material: from Chapter 3, Chan Mainor, Tatiana Segura, and Danny Cardenas, "The Ebola Virus"; from Chapter 5, Christina Straub, "Genetic Variation in the Hawaiian Species *Schiedea* and *alsinidendron*," and Fernando Thadepalli, "Effect of Lomefloxacin, Cefazolin, and Cefonicid on Human Lymphocyte Proliferation," Jamale Carlyle, Dustin Cisneros, and Premal Patel, "Sports Injuries to the Knee," Arturo Pelayo, "The Chemical Study of Resins from Trattinickia apsera used by Coatis for Grooming"; from Chapter 6, Cesar Zamora, for text and graphics from his paper, "Phototoxicity Comparison of Phthalocyanines"; and from Chapter 9, Andy Burchett, "Extended Definition of 'Variable.'"

I thank the organizations for permission to use previously published material: from Chapter 2, R. G. Jahn, "The Persistent Paradox of Psychic Phenomenon: An Engineering Perspective," *IEEE* 70, No. 2 (February 1982); reproduction of plates from George Stubbs, *The Anatomy of the Horse* (New York, Dover Publications, 1976); and thanks also to James Tate, who gave permission to reprint "The Blue Booby," from *The Oblivion Ha-Ha* (Boston: Little, Brown and Company); from Chapter 4, Phuc T. Huynh, "Effects of Food Level and Density of Female Fecundity and Population Stability in *Drosophila Melangaster*," *Journal of Undergraduate Research in the Biological Sciences* 21 (1991); Yung-Chia Melissa Wang, "Summation Bandwidth at Threshold for Gåssler," *Journal of Undergraduate Research in the Biological Sciences* 21 (1991); and Rebecca Torres, "The Biology and Chemistry of Lepidoptera Venoms," *Journal of Undergraduate Research in the Biological Sciences* 21 (1991); from Chapter 9, McGraw-Hill, Inc., for permission to quote from the *McGraw-Hill Dictionary of Scientific and Technical Terms*; Juan Moreno, "Analysis of DNA Synthesis during Acute Polyoma Virus Infection of Mice," *Journal of Undergraduate Research in the Biological Sciences* 21 (1991); and I. Houk and R. Wengler, *Handbook of Applied Hydraulics*, 3d ed. (New York: McGraw-Hill, 1969); from Chapter 10, Harcourt Brace and Company for extensive quotes from S. I. Hayakawa's *Language in Thought and Action*; The University of Chicago Press for the excerpts and drawings Ambroise Paré, *On Monsters and Marvels*, translated by Janis L. Pallister, and to Librairie Droz S. A., for permission to quote from the French-language edition, *Des monstres et des prodiges*, pp. 124, 137, 138, and 141; and descriptive material from *National Audubon Society Field Guide to North American Fishes, Whales and Dolphins* courtesy of Alfred A. Knopf, publisher; from Chapter 11, Barbara J. Stahl, *Vertebrate History: Problem in Evolution* (New York: Dover Publications, 1985); from Chapter 12, William Gilbert, *De Magnete*, translated by P. Fleury Mottelay (New York: Dover Publications, 1958); and, finally, from the Reading Selections, ASM Press for the introductions and translations of "The Germ Theory of Disease" by Girolamo Fracastoro (translation and introduction by T. Brock) and "A *Contagium vivum fluidum* as the Cause of the Mosaic Diseases of Tobacco Leaves" by Martinus W. Beijerinck (translation and introduction by T. Brock), both from *Milestones in Microbiology*, ed. & tr. Thomas Brock (Washington, D.C.: American Society for Microbiology, 1961), pp. 69–85 and 153–56; and "Chronological Note" from *Hippocratic Writings*, ed. G. E. R. Lloyd, tr. J. Chadwick and W. N. Mann (Penguin Classics 1978; first published by Blackwell) copyright © 1950 by J. Chadwick and W. N. Mann.

Thanks to the American Association for the Advancement of Science for permission to reprint the following abstracts from *Science*: from Chapter 4, Douglas C. Wallace, "Mitochondrial Genetics: A Paradigm for Aging and Degenerative Disease?" J. K. Bates, J. P. Bradley, A. Teetsov, C. R. Bradley, M. Buckholzten Brink, "Colloid Formations during Waste Form Reaction: Implications for Nuclear Waste Disposal," W. E. Pickett, H. Krakauer, R. E. Cohen, D. J. Singh, "Fermi Surfaces, Fermi Liquids, and High-Temperature Superconductors," Victor F. Fuchs and Diane M. Reklis, "America's Children: Economic Perspective and Policy Options"; from Chapter 6, M. F. Horan, J. W. Morgan, R. J. Walker, J. N. Grossman, "Rhenium-Osmium Isotope Constraints on the Age of Iron Meteors"; from Chapter 10, Alan Lightman and Owen Gingerich, "When Do Anomalies Begin?"; from Chapter 11, the extended passage from Allan M. Brandt, "The Syphilis Epidemic and Its Relation to AIDS."

I thank the reviewers of my manuscript for their advice and encouragement: George Basile, Stanford University; Dale Flynn, University of California, Davis; Patricia Hartz, University of California, Irvine; Gary Jason, San Diego State University; Luis Villarreal, University of California, Irvine; and Robin White, California State Polytechnic.

I thank Arthur and Nancy Bartlett for their longstanding support and confidence in this project.

Robert A. Varsano, my beloved friend with AIDS, reminded me daily, by his presence, that publishing results is not an academic exercise. His untimely death on July 24, 1995, is commemorated here. In his name, this book is released under the wing of a hope for a better world and with a vote of confidence in the young people who can make it so. The promise of the next generation holds comfort for the last.

*Marilyn Moriarty*
*Assistant Professor of English*
*Hollins College*
*Roanoke, Virginia*
*August 1996*

# Introduction

If you have watched a carpenter, a stone mason, or a sculptor at work, you would have seen that the craftsman studies the medium first. A carpenter studies the wood for the grain and the presence of knotholes to decide whether to cut with the grain or against the grain. A sculptor working with marble studies the stone to assess the strong places and the weak places, the blemishes and the bright spots. No carpenter or sculptor attempts to embody a preconceived idea without knowledge and study of the medium.

We are not accustomed to thinking of language as a medium because, to some extent, language is immaterial: It lacks quantitative substance like weight, color, and mass. While books can be written—and books possess materiality—language is generally thought to be what we *speak* rather than what we see, touch, feel. But the physical immateriality of the medium does not automatically mean that language is without texture.

Language does function as a medium—like wood or stone—but it is an invisible medium. Have you ever wondered, *Why can't I write what I think?* You cannot because the act of writing words on a page is a different operation from thinking thoughts in your head, just as the *concept* of a human figure for a statue is very different from *rendering* a figure in stone.

If you are fortunate to be multilingual or to have studied other languages in school, you should be aware that different languages have different capabilities to create meanings. One language may have a more apt expression for an emotion; another language may have more words for certain concepts. Within and through language we come to shape our experience, for language, ultimately, gives us conceptual categories. So while language does not equal thought [and thus, you cannot automatically write what you think], language *structures* thought.

Consequently, language does not simply give body to pre-existing ideas; ideas come out of manipulating and struggling with the medium, with moving words around, with writing, rewriting, tearing up pages, with having another reader critique your work. Meaning is something you *make*.

The ability to *make words mean precisely* is critical in the sciences. As soon as you generate a hypothesis, that hypothesis takes shape in words. The research question, as a question, is posed in language. When you describe the experiment you conducted, the description is given body in words. Regardless of the intrinsic merit of your research, your results remain virtually meaningless unless they are communicated to others.

*Writing Science through Critical Thinking* is based on two principles. The first principle is this: All writing that you do in the sciences has an audience. Even if you are unable to visualize a specific person's face, you write for and to a reader. We want you to be conscious of that typical reader as you write and, before you write, to consider what that reader could do with the information you impart. Could the reader visualize an object from your description? Could the reader pick that object out of a series of like objects on the basis of your description? Could the reader understand a concept through your explanation?

The targeted audience governs your approach. As a writer, your task is to bridge the information gap between what your audience already knows and what they need to know to understand your point. In the absence of a common vocabulary between the writer and the reader, the obligation lies with the writer—with you—to generate that language. This common language will be slanted for the reader's comprehension. Because you are the writer, the burden of intelligibility rests with you.

The second principle is this: You already possess the skills that will make you a competent writer. Thinking and writing are very similar processes, but thought travels faster than a hand on a keyboard or a pen on the page. When you pick your car out of a parking lot, when you recognize your friend in a crowded room, your brain performs many operations so quickly you are not aware of them. You recognize patterns, discern the variation in the patterns, and then act on that recognition. Every day you make interpretations or explanations about events that happen in your life. *Writing* such explanations requires you to *reconstruct* your pattern of thought and *render* it in words. When you write for a reader, you guide another person through the reasoning processes you undergo automatically in your own mind. Whether you explain how you found your car or conducted an experiment, you make explicit what you have done without conscious thought. Writing serves to transfer automatic or mechanical actions into the realm of conscious thought.

*Writing Science* will show you how to transfer systematic thinking patterns into systematic writing patterns by paying attention to detail. The detail you note in the course of your observations will also be the detail you organize in the course of your writing.

We assume that you can reproduce what you can see and you can "repeat" what you can "hear." While it was not possible to offer many examples for every aspect of this book, it is hoped that the many examples do in fact allow you to learn through the same thought processes involved in scientific experimentation. Through contact with examples—drawn from student work, published papers in journals, or historical classics—you can extrapolate a more general rule (induction), and once you have a rule, you can apply it to other instances to make a determination (deduction). While writing of any type requires work to take control of the medium, we believe that if you can see what it looks like, "hear" what it "sounds" like, then you can learn to write so that it *looks like that*, write so that it *sounds like that*.

Chapter 1, Thinking Science, defines a body of principles and terms to generate a common language. The terminology of induction, deduction, and the scientific method are among those rehearsed. Critical thinking is defined as the ability to differentiate *what you see, what you think you see,* and *what you think it means.* Chapter 2, Writing Science, defines the attributes of scientific writing. You may be surprised to learn that any topic can be appropriate for science writing; there is no intrinsically proper subject of scientific investigation.

Chapter 3, Audience and Purpose, stakes out the shared role of the audience and the writer's purpose in scientific communication. While the concept of writing for an audience may not be new to you, you may not have considered scientific writing in these terms before. The targeted audience influences the kind of language you use, the nature of your explanations, and the type and complexity of your graphics.

Chapter 4, Abstracts, is written from the standpoint of reading abstracts with the underlying goal being your writing them. They are introduced early in the book because they provide a micro-version of the scientific paper (Chapter 6). It is possible to read any kind of writing if you can (1) identify central concepts through key words and (2) identify the relationships expressed in the sentence.

Chapter 5, Using Headings, takes the place of a chapter on organization. It defines and explains the function of headings—titles and subtitles—in scientific papers because the writing events will be organized with respect to discrete sections in the formal paper. Chapters 4 and 5 expose you to information in one domain in (reading, organization) that you will actively put into practice later as you write the formal paper. Chapter 6, The Scientific Paper, describes the parts of the paper, recommends an order of writing events, and directs you to other sections of the book for assistance in structuring your argument. It is complemented by Chapter 7, Graphic Illustration, which guides you through the process of fashioning a graphic and integrating it with the text of your paper.

Chapter 8, Description, teaches you how to create a picture in words. This chapter alerts you to the point that the logic of description follows the logic of writing. Descriptions are useful if your experimentation deals with a special kind of apparatus or a particular kind of subject. In the scientific paper descriptions may also be keyed to rendering or reporting results.

A definition offers the best starting point for many writing endeavors because a definition provides a shared contact point between the writer and the reader. A definition is the writer's way of saying *I start here*, and a reader will recognize the presence of a conscious, reasoned strategy for dealing with the material. Chapter 9 explains how to write and use three kinds of definitions: parenthetical, sentence, and extended. It is hoped that you will learn to use a definition as the foundation of an explanation, an argument, or a discussion, to ensure that you and your reader work from the same starting points.

Classification is both a cognitive act and a writing form. As part of the way that we think about the world, classification enables us to identify the unknown [by comparing it with the known]. Classification, the basis by which you decide *what you think you see*, governs the way you treat the object in question. In addition to explaining the importance of classification, Chapter 10, Classification, extends the applications more generally to provide guidelines for starting a paper, narrowing down a topic, and using the library. Because any essay is also an exercise in classification—with the title functioning as the name of the class and the paragraphs dealing with individual elements in that class—a discussion on the use of the three-part thesis is included in this chapter.

Chapter 11, Comparison/Contrast, explains the fundamental ways that one uses this organizational mode. In particular, this mode is useful in generating explanations for the Discussion section of a formal paper. The applications for Comparison/Contrast are extended in this chapter to include writing the statement of purpose, proposal and recommendation writing, and analysis of results.

Chapter 12, Process, describes three ways to write about events in time: the lab report, the set of instructions, and causal analysis. Appendix I provides an condensed discussion on writing grant proposals; Appendix II abbreviates some important rules for style and mechanics.

The examples in *Writing Science* are drawn from a number of sources. They include a mix of student-written work and classical essays. The essays, included in the Reading Selections, are taken from early scientific writings. These essays model the scientific mind at work classifying or analyzing data. In some instances, the writers worked with limited theoretical models that failed to accommodate their observations. The scenarios created by these examples prepare us for a time when contemporary researchers might be in a parallel situation; the historical scenarios offer hypothetical situations for future scenarios.

# 1

## Thinking Science

When you pick out your car from a crowded parking lot, when you recognize your friend in a crowded room, your mind runs through many processes. You recognize patterns, discern the variation in the pattern, and then act on that recognition. Because you perform such operations automatically, you probably do not pay much attention to the way that you think. You move through an everyday world automatically, seeing, hearing, feeling, tasting, things with all your senses but without registering their specific operation. When you go to the grocery store to buy an apple, you make your selection based on your experience with apples: You look for a ripe one. Perhaps you decide that you will eat it tomorrow rather than today, and your decision becomes more complicated as you select the apple that is not ripe yet but will be ripe later. Picking your car out of a parking lot or picking an apple out of a bunch requires a series of observations, assessments, and evaluations that form the basis of your actions.

Science thinking does not differ in kind from ordinary patterns of thinking; however, science thinking makes ordinary thought processes deliberate, serial, ordered, and focused. The central difference between your everyday encounters with the world and your encounters with phenomena in a scientific setting (or from a scientific perspective) is that, with the latter, your observations are deliberate, a body of received knowledge directs the questions you pose, and a systematic program shapes the answers you seek.

This chapter provides some preliminary explanations to generate a common vocabulary and intellectual context. If the discussions are familiar to you and you are already acquainted with the concepts, press on through the chapter or go directly to the exercises at the end and try your hand at them.

### Patterns and Pattern Recognition

**Pattern**

A pattern is a recurrent motif, event, element, attribute, or phenomenon.

*Recur* means occurring again. The identification of a pattern requires the assertion of resemblance between things, events, or phenomena. If you saw 100 spiral shells arranged in a line on the beach, you would be able to note the similarities and the differences. Common observation reveals, for instance, that some spiral shells twist in a clockwise direction, whereas other shells twist counter-clockwise. If asked to identify the pattern in the following set—[2, 4, 6, 8, 10, 12, 14]—you could identify this as a series of even numbers. Such patterns are routine and common in your everyday experience.

Scientific observation, however, concerns itself with identifying patterns between different *kinds* of phenomena. When Isaac Newton (1642–1727) perceived the dropping of an apple and the orbit of the moon as expressions of the same principle, he identified gravity. The independent English researcher, Richard Christopher Carrington, was the first both to note the occurrence of a solar flare and to connect it with increased electromagnetic activity at the north and south poles: He connected an event on the sun with events on the earth. While the Arab scientist Avicenna (979–1037) was the first to emphasize the connection between water droplets and a rainbow, it was left to the Persian Kamal ad-Din more than 250 years after Avicenna's death to apply the rules of refraction in optics to this phenomenon.

Pattern recognition relies upon the observation of a congruence or resemblance between two or more events, properties or attributes, even when that resemblance seems to stand outside or beyond the obvious range of connections. But the identification of patterns requires more than a moment's intuition for such an observation to receive validation within a specified field of inquiry. Such observations or intuitions require further scrutiny to have lasting value. For instance, you might see a pattern between carrying a rabbit's foot and the grades you get on a test. It may be that every time you carry a rabbit's foot, you get an A on a reading quiz. This resemblance between discrete acts—the grade on a test and the rabbit's foot—leads you to think there might be a pattern.

Is there a connection between these events? And is that connection a *causal* one? In other words, did the rabbit's foot *cause* or *influence* the outcome of your grades? This scenario can be scrutinized more closely through the application of critical thinking.

### Critical Thinking

Starting with the moment you noted the connection, you replay the events in your mind as if they were frames in a movie. In reviewing your thought process, you play the double role of participant and critic. As a critic, in the course of reconsidering the series of events, you start to make some crucial discriminations. You separate the process into three parts: (1) *what you see*, (2) *what you think you see*, and (3) *what you think it means*.

- *What you see* are two discrete events: You observe that (1) you carry a rabbit's root and (2) you take a test successfully. (At the conclusion of a scientific experiment, what you see is documented in the Results section.)
- *What you think you see* is a correlation: When you carry the rabbit's foot, you pass the test; there is a pattern.
- *What you think it means*: A rabbit's foot carries an inherent power to influence good grades or to directly cause good grades. (The *meaning* of the results is elaborated in the Discussion section of the scientific paper.)

But this series of causal relations might be a pattern created in the mind rather than existing in the world. Possibly these events are connected simply because they possess some correlation: What they have in common is you. The agent who pocketed the rabbit's foot is the same agent who took the test. It could have happened to anyone, but it happened to you and you took note of it. Or, alternatively, these events are simply connected by a temporal relation—one event occurred *after* the other—rather than by a causal one.[1] So, while you rightly noted that two events seemed connected, that connection was not a causal one but a coincidental one. To assess precisely the *nature* of this pattern, you need to devise a way to determine if *what you see* and *what you think you see* actually possess a correlation. In other words, you need to test the congruence between the pattern you noted in your mind and a possible pattern in the world.

How could you test the connection between the grade and the good-luck charm?

To reduplicate your results under less random conditions, you might run trials by taking a quiz without the rabbit's foot or by giving the rabbit's foot to all your friends who will take examinations. In so doing, you subject the pattern of perceived causal relations to testing. You might, in short, follow a systematic pattern of experimentation to subject your intuitions or perceptions to further scrutiny. This systematic pattern would enable you to determine the extent to which your intuition or thought had actual validation in the world. It would enable you to distinguish between what you see (observation), what you think you see (pattern), and what you think it means (significance).

When you assess the extent to which something that occurs in your mind occurs in the world, you make the shift from a *subjective* approach to an *objective* approach.

## Subjectivity, Objectivity, and Consensus

*Subjective* refers to your individual perceptions, ideas, intuitions—categories of events that are specific to an individual or to an individual's point of view. Subjectivity derives from your experience, background, thoughts, likes, dislikes; it is the essence of the personal. Objective refers to events in the world as they exist independently of an individual point of view.

Another word for "the world as it is" is *empirical*. Empirical reality is a different category from subjective reality. While one's feelings or emotions possess a sense of authenticity and truthfulness to an individual, such subjective feelings cannot necessarily extend to all other individuals. Sometimes a subjective intuition may coincide with empirical evidence; sometimes it may not. Individuals may reach agreement or consensus on matters not necessarily true; in other instances consensus may correlate with objective truth. These distinctions are important because they clarify a language that enables us to talk about truth and experience.

The language of truth differs from the truth itself. Truth may be defined as the ontological status of a law or phenomenon that can exist independently of our knowledge of it. Language about the truth attempts to make statements with a high degree of correlation with empirical reality. A truthful statement is characterized by corroboration with the facts. Henceforth, *truth* will be understood in its stipulated meaning as correspondence with the facts.

If another person works on the same research project as you and arrives at the same results, your results are said to *corroborate* or support each other. If a substantial number of individuals arrive at the same agreement, they may be said to have arrived at a *consensus*.

*Consensus* means that a sufficiently large group of individuals agrees with you to the point that general accord exists; this agreement may stem from shared recognition of a fact in the world or it may stem from shared opinions. *Consensus* is not synonymous with *truth*; *consensus* is an agreement word.

Over a long period of the history of science, individuals have come to a consensus over systematic ways to test the connection between intuition and fact, to determine whether a perception or idea can be validated by empirical results. While not a standardized procedure, this method goes by the name of the scientific method. The scientific method does not lay claims to strip away false appearances to arrive at the absolute metaphysical truth; rather, it is based on systematic thinking patterns that attempt to reveal or explain causal relations between substances, events, or phenomena.

## The Scientific Method

*The scientific method* is the term given to an approach to investigation that is based on empirical evidence. *Empirical* refers to *demonstrated evidence* rather than purely theoretical speculations or explanations based on faith. While all scientific disciplines acknowledge in some form or another that they employ a standard method of investigation, variations exist within the disciplines in terms of the way this standard procedure operates.

The scientific method is not A method or THE method; in general, scientific method refers to a consistent and logical manner of framing questions about the world and a systematic way of finding answers to those questions.

While we usually think of experimentation as the crux of the scientific method, it should be noted that not all kinds of inquiry lend themselves to testing under controlled conditions. While the valorized testing norm in the physical and biological sciences has been the experiment, studies in astronomy, oceanography, meteorology, and paleontology do not always present situations subject to controlled conditions and repeated tests;[2] such studies do make use of observation, inference, and comparative analysis—methodologies that specify the terms of their validation. Similarly, mathematical calculations may not find direct corroboration through empirical evidence; the formulation and assessment of such systems exist in a domain that differs from that of empirical sciences by the absence of a physical referent. Such systems are validated by the strictness of measurement imposed on precise relations in that domain.

The scientific method draws upon the studied observation of phenomena, the calculated intervention in phenomena in order to produce an answer to a question, and the evaluation of the answer to the question in light of both the question that is asked and the way it is assessed. These areas can be broken down into three parts: observation, intervention, and interpretation.

**Observation**

Observation is a word and a process usually associated with the sense of sight. To observe is to look. But if we press the word, *observe* really means to take note of, notice, pay attention to. The word *observe* tends to be used in two senses, although the dual meanings are slightly obscured. *Observe* describes cases of casual looking and cases of intense, focused, attention-paying looking. Running through a park on a spring day, a jogger might observe a cloud pass before the sun. This observation is casual, the jogger not inclined to make much of the fact. However, a meteorologist studying cloud patterns and wind streams might observe—not simply to note but to *mark*—the way a cloud passed before the sun. This observation carries with it a purposeful looking, a looking-in-order-to-mark-or-note. In this case, the observer sees *for the purpose of* observing.

Observations need not be purely visual; one might take note of the world through other senses. Observations can also be made through the mediation of instruments that enhance the human range of sense perception. Nor are observations restricted to natural phenomena; some events contrived in a laboratory may never take place in nature.

## Direct Phenomena versus Effects

*Direct phenomena* refers to phenomena encountered through direct experience, although mediated by an instrument. A microscope, while a mediating instrument, enables us to see the form and texture of a minute organism. Instrumentation enables us to see an organism better or to hear a tone clearer; instrumentation thus enhances the range of human perception. A direct phenomenon is considered to be the phenomenon or object subjected to the range of instrumental enhancement. Some degree of attention must be paid to what one observes via the instrument.

But a realm of available data exists beyond the range of human sense perception or its enhancements. Some instruments measure the *effects* of phenomena that cannot be sensed directly or through enhancing devices. Atoms, too small to be seen by the naked eye or high-power microscopes, can be detected by instruments that measure the effects of atoms through energy release. An X-ray photograph does not show your bones. It shows the *effects* of radiation as it is absorbed by bone and recorded on a photographic plate. How does a geiger counter measure radiation? It does not measure radiation directly; instead, a geiger counter records the frequency of a charge track through a gas field that leads to a connection between a cathode and an anode.

Phenomena observed directly or through their effects are considered to be in the realm of observation.

## Possible Sources of Error in Observation

Two possible sources of error in observation come from mediations—those pertaining to instruments and those pertaining to the observer. The *object* of observation (the phenomena under scrutiny) is, to varying degrees, influenced by the *means* of observation. What you observe—direct phenomena or the effects of phenomena—reflects the capabilities and the limitations of the instrument. The instrument by which one measures or observes phenomena offers a possible source of error in the collection or evaluation of data. (Thus the need for an assessment of results as they may be influenced, created, or limited by the very question posed and the means used to arrive at answers.) Even the naked eye is an instrument that has limitations governed by the laws of optics.

As a careful scientist, you need to be able to differentiate between *what you see* from *what you think you see*. Do not confuse your perception of a thing with your perception of its representation. Some instrumentation does not show *exactly* what takes place; rather it *represents* what takes place.

Sometimes a distortion may come about through our physical and temporal perspective on events. In the passage of a day, as the sun rises in the east and sets in the west, we think we see the sun move; a viewer positioned at a point outside the sphere of the earth would understand that the movement of the sun is an effect of the earth's rotation. Our sense of the sun's movement is produced by our position on the earth.

In addition to the limitations of perspective and the possible distortion through representation, there is a source of mediation that is not physical but cognitive. In an attempt to regularize or make sense of what we think we see, it is possible that the information from a preconceived idea may override the immediate perception of the object before our eyes. In *The Structure of Scientific Revolutions* Thomas Kuhn argues that a body of received ideas (what he calls a *paradigm*) comprises a structure within which we make sense of discrete data. Norwood Russell Hanson states that "an important factor in how we see the world is the knowledge and experience we bring with us to observational situations. Our knowledge, our experience, and . . . our *language* are, as it were, spectacles that we wear behind our eyes."[3]

Experience, knowledge, and preconceived beliefs influence our expectation in ways imperceptible to our conscious thought, and they condition our expectations in ways we do not always know. Medieval European anatomy was hindered for centuries because scientists permitted their expectations, drawn from the work of classical philosophers, to govern their perception; some medieval anatomists failed to see the veins before their eyes even when they observed a dissected body. The self-fulfilling power of expectation applies to recent observations as well. When Charles Doolittle Walcott studied the reservoir of prehistoric fauna in the Burgess shale (Canadian Rockies) in the early twentieth century, received taxonomic views dominated his classification of new fauna; he tried to put new species into old ready-made boxes. Subsequently, other researchers recast those old systems. These sometimes-unconscious interpretations occur because a scientist's expectation of what *should be* may constrain *what is*. While the history of sciences enables us to recognize the operations of conceptual paradigms in earlier times, it is more difficult to assess how contemporary received ideas shape our views. Instrumental mediations are easier to gauge than conceptual mediations because the physical fact of the device calls our attention to its instrumental role. Conceptual mediations—the way our preconceived ideas operate—are harder to assess because they are not physical or sensual; they are cognitive, inherent in the structure of thought and language. A sense of history, or an understanding of your perspective as the endpoint of a trajectory of ideas, will help you achieve a critical evaluation of your point of view. That said, we ultimately have no way of knowing how our scientific perspective in the twentieth century will bear up under the scrutiny of scientists to come, generations hence.

If we stay in the realm of casual observation or of taking note of everyday life, we can be naturalists or poets or simply observant people. What makes a scientist different, though, is that directed observation is tied to an active engagement in natural or contrived process. For example, in the eighteenth century, the Italian physiologist, Luigi Galvani was studying muscle contraction in frogs legs. In the course of his research, casual observations led to directed ones. Galvani describes the process as the focus of his seeing changed:

> Therefore having noticed that frog preparation which hung by copper hooks from the iron railing surrounding a balcony of our house contracted not only during thunder storms but also in fine weather, I decided to determine whether or not these actions were due to the action of the atmospheric electricity. . . . Finally . . . I began to scrape and press the hook fastened to the back bone against the iron railing to see whether by such procedure contractions might be excited, and whether instead of an altera- tion in the condition of the atmospheric electricity, some other changes might be effective. I then noticed frequent contractions, none of which depended on variations in the weather.[4]

For Galvani, casual observation preceded directed observation; di- rected observation eventually led to his formulating a hypothesis and to testing it by constructing a situation that would enable his isolation of the essential conditions under which the phenomenon took place.

Engagement in the process shifts a scientist's mode of seeing by placing observation within the field of causality. The movement of a cloud or the quivering of a limb ceases to figure as an innocent or contingent event; it suddenly comes under scrutiny as an act that is located within a field of causes. When observation becomes focused (as one sees purposefully), the nature of observation changes. And as observation becomes directed, sci- entists attempt to carve out and determine the conditions under which such observations occurred. Focused, attention-paying observation is the way one sees when doing science. *Doing* science requires a conscious, focused, and systematic interaction with natural or contrived processes.

## Intervention

Experimentation means a try or a test; it is the primary means by which an observer intervenes in a natural or artificial process. Experimentation refers to the act of testing a phenomenon to see if it can be repeated or replicated. While grounded in observation, experimentation makes the scientist an active participant in manipulating phenomena.

Scientific method holds as a convention the regularization of the con- ditions of experimentation, so that, under controlled conditions, the vari- ation in one variable or in a limited set of variables can be documented. A *control*—an analogous experiment in which the variable is not manipulated—is run alongside the experiment to serve as a basis of comparison and contrast.

But direct manipulation of physical substances under controlled conditions is not the only means of scientific intervention. Conceptual or cognitive processes used to describe or explain events or phenomenon that can not be investigated under controlled conditions may be considered mental interventions. Mathematical modeling, classification, and comparative analysis may be considered in this light; taxonomy is a mental, not a physical, apparatus. Taxonomic categories do not exist in nature; human beings impose those categories upon flora and fauna to describe patterns of relations. No chimpanzee conceived itself as a chimpanzee or a primate, taxonomically related to human beings. Similarly, some mathematical models used to process raw numerical data, while possessing a capacity for making predictions, do not require empirical verification to validate their conclusions. Some fields of research like oceanography or meteorology study bodies beyond human control. Paleontologists find the biological definition of a species—the criterion of interbreeding—to be impractical when dealing with species no longer extant. In such cases, the impossibility of specifying controlled conditions neither disqualifies these fields from the arena of scientific investigation nor undercuts the strength of arguments proposed through logical, mathematical, or comparative means.

### The Research Question: Getting the Idea

A purpose guides the creation of an experiment or observation. Experiments are devised to answer questions: "What happens if . . . ?" or "Why?" Why, asked William Harvey, are there valves in veins and not in arteries? In the formulation of his germ theory of disease (see the appendix), Fracastoro asked, "What, shall we say, is the nature of this infection? Is it a corruption of these particles [infected or putrefying parts] or only an alteration? What, in short, happens to those particles?" In the mid-thirteenth century, Frederick II, the Emperor of the Holy Roman Empire sent questions to the finest Muslim scholars in North Africa and the Middle East. To Al-Kamil, sultan of Egypt, Frederick wrote, "Why does a stick partly immersed in water appear to be bent?" Whether this answer was among the mathematical explanations Al-Kamil returned is not known, but at one point the sultan sent an astronomer to the emperor's court in Italy. The German physicist, Wilhelm Conrad Roentgen, carrying out an experiment with cathode rays in a Crookes tube, noticed that a fluorescent screen nearby became lit when the current passed through the tube. Why, he must have wondered to himself, did a sheet of metal over the tube, but not a sheet of black paper, prevent this fluorescence? Why, asked Edward Lorenz in the early 1960s, did a small numerical error fed into his computer model of weather patterns produce such a radical variation in the output? The search for an answer to that question led to the mathematical equations that marked the advent of chaos theory.[5]

The research question generally proceeds from observations. It shapes the answer and the means to investigating it. The question narrows and specifies the nature of the inquiry; it marks out relevant observations that merit notation. At this point in the process, no hypothesis has been formulated. The question emerges as a response to phenomena.

## The Hypothesis

A hypothesis does not emerge from out of thin air. It is formulated as a possible answer to the question. A hypothesis can be an educated guess about the possible answers to the question or a supposition about anticipated results. In order to devise a test—to select measuring devices, substances, organisms, time frame, computer access—a reasonable sense of the way results *should/might/could* turn out is framed. To formulate an educated guess or a supposition one must have some familiarity about the subject to inform the question that is posed.

## Forming the Hypothesis

Let us say you have a piece of copper, and by accident you run an electric current through it. You observe that a property of copper is that it conducts electricity. A friend brings you a piece of copper from North Dakota, Tierra del Fuego, and Mongolia; these separate pieces of copper conduct electricity. You eventually make 600 repeated trials with different samples.[6]

Because it would be fruitless to go about collecting all the copper in the world to test for conductivity, at a certain point your approach shifts from collecting and observing discrete instances to making a categorical statement about them. By generalizing from the data you have heretofore collected, it is possible to extrapolate a rule about conductivity in copper, and so, on the basis of your contact with a number of discrete instances, you identify a *pattern*, that takes the form of a general law[7]: all copper conducts electricity. This is a reasonable *inductive* inference based on a large sampling and repeated trials.

This logical process that reasons from particular instances to a generalization about those instances is called *induction*.

The induction is made at the point of generalization.

After you have formed this general rule or law about the property of copper, a geologist gives you a piece of copper taken from 500 feet below the surface of the South Pole. Based on your previous work and your formulation of the law, you have reason to believe that this piece of copper, like the others, will conduct electricity. You *hypothesize*, on the basis of previous samples and the generalization derived from them, that this instance will support your generalization.

The concluding statement in your induction—all copper conducts electricity—becomes the starting point for a chain of *deductive* reasoning.

You hypothesize that, given the general principle that copper conducts electricity, this piece will behave in the same way. When you test this new sample against the generalization, you find that this piece too conducts electricity. This hypothesis has been deductively validated.

The process that tests a particular case against a general law is called *deduction*.

*Induction* reasons from discrete examples to general principles.

*Deduction* reasons from general principles to particular instances.

In both deductive and inductive operations, beginning statements (premises) were construed as evidence for the conclusion. In the inductive procedure, we argued that

Copper 1 conducts electricity.
Copper 2 conducts electricity.
Copper 3–600 conducts electricity.
>    Therefore, all copper conducts electricity.

In the deductive scenario, we argued that

All copper conducts electricity.
This piece is copper.
>    Therefore, this piece conducts electricity.

Because not every piece of copper in the world was tested for the induction—although a generous balance of representative samples and repeated testing was done—the possibility remains that one instance could subvert the conclusion. Nonetheless, this induction, while not absolutely conclusive, possesses a high degree of *probability*.

However, a deductive proof offers a stronger evidentiary relation than an inductive one. Inductions are assessed in terms of probability with relative degrees of strength: more probable, less probable. A deduction, however, is not assessed in terms of probability. Its conclusion follows *absolutely*, not probabilistically, from its conclusions.[8] Thus, deductions tend to provide stronger arguments than inductions.[9]

Note that in deriving your inductive principle, your generalization was phrased in universal terms: *all* copper. A universal statement about all copper needs only one instance to disconfirm it. However, if the hypothesis had been phrased "*Some* copper conducts electricity," all cases that *did* conduct electricity would confirm the hypothesis, and all cases that *did not* would likewise confirm the hypothesis.

Note also that the difference between inductive reasoning and deductive reasoning rests primarily at the point which one enters the process.

To continue with the scenario of the conductive properties of copper, let us elaborate the argument by adding a possible complication.

What if you encountered a bundle of wires among which was a copper wire. You reason that this bundle should conduct electricity, but you find, upon testing, that it did not.

What would you do then?

Logically, you could re-examine your previous reasoning and experimentation by reconsidering:

(1) *experimental procedure* (Did you test this piece in the same *manner* that you tested all the others?);

(2) *experimental apparatus* (Did you conduct this test with the same *tools* you used for the other tests?);

(3) *experimental object* (Was this sample really copper? Did this piece of copper differ from other the pieces in any way?);

(4) *generalization* you formed by induction (Was this generalization logically formulated by you or by your intellectual predecessors? Was there an error in its formulation? Did your experiment expose sources of error in the formulation of your generalization?).

Provided that your experiment was not subject to logical or methodological error, you could derive a new research question: What conditions are necessary for electricity to be conducted through copper? The entire procedure of testing would then be replicated with the new question.

Another form of reasoning that enables the formation of a hypothesis rests on the use of analogies. An *analogy* asserts that one object, event, or phenomenon is like another object, event, or phenomenon. For instance, the assertion that the immune system functions like an army proposes a comparison between the human body's defensive force and the body politic's defensive force. To open up an analogy, one investigates the similarities and differences between the two elements in the analogy: In what manner is the immune system like an army? In what manner is the immune system not like an army? One must go further of course in exploring the implications of such an analogy.

A common misconception holds that science proceeds entirely by purely inductive means. Strictly speaking, the greatest limitation to the inductive view is that one starts with no preconceived ideas; this was the view as it was promulgated by Francis Bacon (1561–1626). Having no preconceived ideas about copper, you randomly decided to run a current through it. While you might dedicate your life to testing the conductivity of every piece of copper, life-long dedication would not necessarily lead to any greater insight into how or why copper conducts electricity. No scientist works from a purely inductive approach. Most would find undirected testing or experimentation to be pointless and potentially meaningless. Charles Darwin ridiculed pure induction, stating that anyone who believed in induction "might as well go into a gravel pit and count the pebbles and describe the colours."[10] A reason for investigating the conductive properties of copper puts this experimental action in the realm of meaningful activities.

In contemporary scientific parlance, the inductive method, as it was framed by Francis Bacon to mean the extrapolation of a rule from contact with discrete data collected with no preconceptions about the results, has fallen into disrepute and been revised. Contemporary theories of science maintain that scientific reasoning follows from a combination of inductive and deductive reasoning, based on the formulation of a hypothesis framed from within a context of knowledge, against which separate cases are tested (deduction); this theory goes by the name hypothetico-deductive method, to indicate that a reasonable speculation about what may emerge is included in the process.

## Assessing the Hypothesis

To test a hypothesis means to subject it to trial. Positive evidence produced by testing confirms the hypothesis; negative results produced by the test disconfirm the hypothesis. Hypotheses may be partitioned into auxiliary hypotheses, each of which contributes to confirmation of the primary hypothesis. For example, when William Harvey proposed that the blood in the human body circulates, he also assumed that the circulating blood was the same blood throughout. (See Exercise 2 at the end of this chapter.)

A hypothesis functions in two domains: (1) it is a statement made in language and (2) it is an assertion about the world. The domain delimits the kind of evidence used to confirm or disconfirm the hypothesis. (1) The hypothesis, considered as a speculation framed in language, can be validated by other language statements. Statistical models, for example, frame hypotheses in a mathematical language; validation is derived from within that domain. Because validity is ascertained by the precise delineation of relative values, the hypothesis cannot be tested by empirical evidence because it is not constructed within that domain. (2) As a speculation about events in the world, a hypothesis can be validated by empirical evidence. However, a hypothesis must be phrased in such a way as to render it testable. For example, if on the basis of anecdotal reports, you hypothesize that Bigfoot exists, you would require empirical evidence to verify your hypothesis. The absence of empirical evidence for Bigfoot, however, would make this hypothesis untestable, whereas a hypothesis about the *reports* would be testable.

Some hypotheses can be validated by both empirical evidence and language statements.

In scientific inquiry, the intent is NOT to prove that the hypothesis is true, for, if your goal is to prove a hypothesis is true, it is possible to structure an experiment such that it will yield the desired results. In such a case, it is possible to produce repeated number of verifications.

*The object of experimentation is to falsify the hypothesis.*

If you test a hypothesis with the intention of proving that it is false, you will apply more severe and rigorous tests. If a hypothesis passes many severe tests that attempt to show that it is false, that hypothesis may be said to be corroborated or verified.

*Proving* a hypothesis proceeds from the assumption that the goal of the experiment is to demonstrate that the hypothesis is false; a hypothesis that is not falsifiable is verified.

The hypothesis must be formulated in a way that provides the best answer to the research question. To revisit Luigi Galvani (above): after Galvani had induced contractions in a frog's leg, he continued to test for the source in electricity in the animal tissue. Galvani hypothesized that electricity was a property of animal substance. Although his interest lay in understanding the nature of electricity, Galvani was a physiologist, and his hypothesis directed him to a series of experiments that explored animal physiology. It was left to Volta, who came upon an off-print of Galvani's work, to hypothesize more accurately that electricity was an inorganic phenomenon rather than a quality inherent in animal matter. Thus, Galvani's hypothesis limited, not extended, his investigation.

One must also consider the value of a hypothesis that supports already demonstrated physical laws. While it may possess the merits of testability (falsifiability), it would not shed any new light on the subject.

A hypothesis may be assessed at four different levels.[11]

(1) *Logical consistency.* Is the hypothesis logically consistent and not self-contradictory? Is it framed in the context of a logical approach? Because hypotheses grow out of a received body of knowledge and expectations about what should happen, does this hypothesis reflect the most adequate statement or response to the research question?

(2) *Explanatory value.* Does the hypothesis promise to offer some explanatory value by making the observed phenomena intelligible? In other words, does the hypothesis promise to clarify, explain, or suggest an interpretation? *What* can the hypothesis explain? What advantage does this hypothesis offer over other possible theories? What are the reasons for discarding alternative hypotheses?

(3) *Relation to other theories.* How does the hypothesis compare with other hypotheses or theories current in the field? Does it merit testing? Has it been tested before? Does it represent an advance or a new insight with respect to alternative theories? (Inconsistency with other theories is no reason to reject it.) Given two hypothesis of equal explanatory value, which

is more simple—i.e., which contains the fewer irreducible assumptions?[12]

(4) *Empirical demonstration.* Does empirical evidence bear out the hypothesis? *Can* empirical evidence bear it out or must validation be found another way? Is the hypothesis *capable* of being tested (falsified)?

## Setting Up the Test

An experiment is expected to reveal causal relations. To assess causal relations, controlled conditions are stipulated to limit the number and function of independent variables in order to focus on the ones under investigation.

To understand how a process works, one generally attempts to learn how to make it stop. Control of the variables that prevent a process enable a definition of the variables necessary to produce the process. If it is assumed that $A$ produces $B$, the question can be formed: "Under what conditions will $B$ fail to be produced?" If the absence of $A$ correlates with the absence of $B$, then the empirical results confirm the hypothesis. But in order to apply rigorous tests to the hypothesized causal relation—to attempt to falsify it—it must be demonstrated that $A$ and only $A$ causes $B$, that no $C$, $D$, or $E$ may cause $B$. Consequently, to achieve a higher degree of confirmation, one generally arrives at definitive results by a series of progressively refining experiments.

## Reviewing the Results

Given that a certain body of data has been derived from the experiment, how does one assess the data? One may assess the data in light of methodological soundness. Repeated tests should yield a certain consistency in results.

Because the data have been derived as the response to a research question, the data can be explicated in terms of the extent to which they confirm or disconfirm the hypothesis. The data may be submitted to the following questions:

(1) Does the evidence confirm or disconfirm the entire hypothesis? Does it confirm or disconfirm part of the hypothesis? How does confirmation or disconfirmation of an auxiliary hypothesis confirm or disconfirm the main hypothesis?

(2) Do the results change the status of the hypothesis?

(3) To what extent may be the results be generalized?

Assuming that you have performed the experiment and have decided that your hypothesis possessed explanatory value, that the experiment offered the most appropriate means to test that hypothesis, that the results were derived from methodical processes which were both theoretically and methodologically sound, how do you explain what the results mean?

## Interpretation

Earlier in this chapter, a distinction was made between *what you see, what you think you see,* and *what you think it means.* Meaning refers to the way that facts are interpreted in light of a pattern. An explanation may be considered as the narrative constructed around the facts to make the facts cohere. A narrative need not be a fiction but it does tell a story of causal relations: how and why an event took place within a situation. Explanations are constructed within the context of models or paradigms.

Paradigms offer a context within which a research question and a hypothesis are formulated and within which the data derive their significance. A twentieth-century mathematician studies or thinks about light in the context of Einsteinian physics; a twelfth-century mathematician studied or thought about light in the context of simple optical laws. Data will mean different things to each.

Sometimes new data or findings bring about changes in scientific thought. For instance, innovation in materials science enabled consequent changes in engineering. More often it is the case that change does not come about through new data; change comes about through *new interpretation* of the data. Thomas Kuhn, in *The Structure of Scientific Revolutions,* noted that revolutionary changes in science emerge when traditional paradigms fail to explain all the data leading to a questioning the theory or the natural law that underlies empirical investigation. Similarly, Karl Popper maintained that theory of "refutation followed by revolutionary and yet conservative reconstruction"[13] offered the best explanation of the history of scientific change. This is to say, conceptual paradigms can change when their explanatory value decreases or when the model becomes too elaborate to accommodate all the facts. Discoveries need not arise from the discovery of new data; they may arrive from new interpretations that enable us to see old facts in a new way.

### Exercises

**1. Galileo's "Starry Messenger"**
In "The Starry Messenger"[14] (1610) the astronomer Galileo Galilei asks the following question:

> If that part of the lunar surface which reflects sunlight more brightly is full of chasms

(that is, of countless prominences and hollows), why is it that the western edge of the waxing moon, the eastern edge of the waning moon, and the entire periphery of the full moon are not seen to be uneven, rough, and wavy?

Instead, the moon appears "precisely round as if they (the edges of the full, waning, and waxing moon) were drawn by compass." If the moon's observable gullies and valleys were widespread over the surface, extending even to the edge of the moon, the moon should appear to us almost like a "toothed wheel." But the moon does not appear in this way. Galileo offers an explanation.

> . . . on earth the summits of several mountains close together appear to be situated in one plane if the spectator is a long way off and is placed on an equal elevation. Similarly, in a rough sea the tops of waves seem to lie in one plane, though between one high crest and another there are many gulfs . . . . Now since there are many chains of mountains and chasms on the moon in addition to its periphery, and since the eye, regarding them from a great distance, lies nearly in the plane of their summits, no one need wonder that they appear as arranged in a regular and unbroken line.[15]

a. Make a list of Galileo's observations. What does he see?

b. What does Galileo *think* he sees?

c. How does Galileo explain his observations?

d. What alternative explanations might account for his observations?

e. Astronomer Joe Blow suggested that the edge of the moon appears smooth because it *is* smooth and that the gullies and mountains are located only on the non-edge surfaces. Is there any information in Galileo's observations that would refute Joe Blow?

f. What analogy underlies Galileo's explanation?

## 2. William Harvey and the Circulation of Blood

William Harvey (1578–1657), anatomist and physician to the King Charles I (who gave Harvey access to his deer park), was famous for his discovery of the circulation of blood.

Previously, it was believed that the body consumed blood as fast as the liver produced it; the liver produced blood continuously. Blood traveled the following path: Produced in the liver, venous blood carried nutrients from digestion to the body. A small amount of venous blood that was not consumed by the body moved into the heart, and from there "life-giving" blood (a combination of blood and refined air/spirit) went out from the heart and was absorbed by the body.

Andrew Wear summarizes Harvey's contribution: "Harvey made quantitative experiments which showed that far too much blood left the heart in a given time for it to have been used up by the body and replaced by the blood manufactured in the liver. . . . [Harvey's] quantitative argument suggested that the blood must flow continuously in a circle, otherwise the body would burst."[16]

Without the newly invented microscope (which would have enabled Harvey to identify capillaries, those structures invisible to the naked eye that connect veins and arteries), Harvey logically reasoned that the two systems of blood were connected and circulated the same blood. How could Harvey demonstrate circulation?

> Now make a test on a man's arm by applying such a bandage as they use in

blood-letting. . . . from these facts any reasonably careful observer can readily learn that the blood enters through the arteries, for a tight ligature of those vessels does not *draw** anything; the hand keeps its color, nothing flows into it, and it is not distended. When, however, the arteries are freed a little (as they are on a change to a medium tight ligature) it is clear that blood is being driven into them forcibly . . . and in adequate amount, and that the hand is swelling up. Where the arteries pulse of course, the blood flows on, as it does; but where they are pulseless, as in a tight ligature, there is no flow save above that ligature. Meanwhile, there can be no inflow through the veins when they are compressed, and this is indicated by two facts (i) they are much more swollen below the ligature than they are above it or are wont to be after its removal, and (ii) that when compressed they supply nothing to the parts above. So obviously the ligature prevents the return of blood through the veins to the parts above, and causes the veins below itself to remain swollen. [17]

a. Given the Galenic model of blood production and function, Harvey's hypothesis that the blood circulated generated auxiliary hypotheses. Using your own reasoning faculties, generate one or more auxiliary hypotheses.

b. Read through Harvey's description and differentiate the following: what he sees, what he thinks he sees, what he thinks it means.

c. Does this experiment alone confirm (or not) Harvey's hypothesis about circulation? Does it confirm a part of his hypothesis?

d Would alternative explanations account for the results of his experiment?

e. Harvey closes chapter nine of *The Circulation of the Blood* with the observation that a butcher will not be able to bleed an ox completely (after stunning it with a blow to the head), unless he cuts its throat before the heart stops beating. Does this observation support or undercut Harvey's hypothesis about the circulation of the blood? Does this observation support or refute a part of that hypothesis?

f. Harvey's hypothesis was spurred by the quantitative implausibility of diffusing a regular volume of liquid through the body. What does the butcher's practice of bleeding an ox while alive contribute to the quantitative element?

g. What does this instance with Harvey show about the relation between theory and quantitative data?

h. Compare and contrast Harvey and Galileo in terms of their methodology. How are they alike? How are they different?

### 3. Galvani's Observations

Luigi Galvani's description of the way he began to think about electricity appears earlier in Chapter 1. Go through this excerpt and identify his hypothesis, his observations, and his experiment. What is the difference between the observations in sentence (1) and (3)?

---

*   According to the Galenic view, nutritive blood was attracted or drawn to parts of the body from the liver.

(1) Therefore having noticed that frog preparation which hung by copper hooks from the iron railing surrounding a balcony of our house contracted not only during thunder storms but also in fine weather, I decided to determine whether or not these actions were due to the action of the atmospheric electricity.... (2) Finally ... I began to scrape and press the hook fastened to the back bone against the iron railing to see whether by such procedure contractions might be excited, and whether instead of an alteration in the condition of the atmospheric electricity, some other changes might be effective. (3) I then noticed frequent contractions, none of which depended on variations in the weather.[18]

### 4. Henry Gosse and Adam's Navel

Considered the finest British naturalist of his day, Patrick Henry Gosse (1810–1888)[19] wrote many books on plants and animals, published technical papers on marine invertebrates, and lectured widely. A committed empiricist and a committed fundamentalist of the Brethren Plymouth Church, he published *Omphalos: An Attempt to Untie the Geological Knot* two years before Darwin published *The Origin of Species*; this work explains the evidence presented by geological strata and fossils (with which, as a naturalist, he was exquisitely familiar) in a way that was consistent with his creationist point of view.

Gosse's thinking might be paraphrased as follows: Why did Adam have a navel? If Adam were created, rather than born, he should have no navel. Gosse's explanation for Adam's navel could be extended to account for fossil remains: God made Adam with a navel just as God made the earth with fossils. Adam's navel was intended to enable him to fit in with subsequent generations of humans that were born not created. Similarly, prehistoric animals were not created as if they were "brand new"; they were created in their "used" condition to provide a transition to a past. Like Adam's navel, the geologic record was created as remains. Old dinosaurs were created old; some dinosaurs were created with food in their bellies, others as embryos in the egg. Even coprolites demonstrated the completeness of this plan.

    a. Assume that you share Gosse's perspective on dinosaur remains, and differentiate what he saw, what he thought he saw, what he thought it meant.

    b. Working with Gosse's premises as your starting point, formulate a hypothesis and devise an experiment that could be used to test it.

    c. On what *scientific grounds* can Gosse's theory be validated or invalidated?

### 5. Take This Fish

Turn to the Reading Selections at the back of the book and read Samuel H. Scudder, "Take This Fish and Look at It." What are the central points about observation? How does observation relate to description? Was Professor Agassiz directing Scudder toward inductive or deductive reasoning? Explain.

### 6. Tobacco Mosaic Virus

Turn to the Reading Selections and read Martinus W. Beijerinck's "A *Contagium vivum fluidum* as the cause of the mosaic diseases of tobacco plants." Answer the questions at the end of the reading.

## 7. Modeling DNA

Turn to the Reading Selections and read F. H. Crick and J. D. Watson's seminal paper, "A Structure for Deoxyribose Nucleic Acid." Answer the questions at the end of the reading.

## 8. Syllogisms in Wonderland

In the previous discussion of induction and deduction, you were presented with a syllogism. A syllogism is a logical pattern of reasoning that proceeds as follows: a=b, b=c; therefore a=c.[20] It is possible to provide a strong argument by reasoning through syllogisms.

All copper conducts electricity.
This piece is copper.
>    Therefore, this piece conducts electricity.

C. L. Dodgson was a British mathematician who wrote two books on logic. (Dodgson is better known as Lewis Carroll, the author of *Alice in Wonderland*.) The following syllogism is adapted from syllogisms taken from his book *Symbolic Logic*.

All chickens speak French.
This creature is a chicken.
>    Therefore, this creature speaks French.

Compare Dodgson's syllogism to the copper syllogism. Do you accept the conclusions of the copper syllogism? Of Dodgson's? Why or why not? (You will encounter a further explanation of Dodgson in the following chapter.)

## References

Crimp, Douglas. *AIDS: Cultural Analysis, Cultural Activism*. MIT Press, 1988.

Dobzhansky, T., F. J. Ayala, G. L. Stebbins, and J. W. Valentine. *Evolution*. W. H. Freeman and Company, 1977.

Dodgson, Charles. *Symbolic Logic*. Macmillan, 1896.

Galilei, Galileo. *Discoveries and Opinions of Galileo*. Doubleday, 1957.

Gleick, James. *Chaos: Making a New Science*. Penguin Books, 1988.

Gould, Stephen Jay. *Wonderful Life: The Burgess Shale and the Nature of History*. W. W. Norton, 1989.

_____. *The Flamingo's Smile*. W. W. Norton, 1985.

Hanson, Norwood Russell. *Perception and Discovery: An Introduction to Scientific Inquiry*. Freeman, Cooper, & Company, 1969.

Hartz, Patricia. "An Interdisciplinary Guide to Critical Thinking." Manuscript in process. 1995.

Harvey, William. *The Circulation of the Blood and Other Writings*. Translated by Kenneth Franklin. J. M. Dent & Sons Ltd, 1990.

Jason, Gary. *Introduction to Logic*. Jones and Bartlett, 1994.

Kuhn, Thomas S. *The Structure of Scientific Revolutions*, 2d ed. University of Chicago Press, 1962, 1970.

Lewis, Bernard, ed. *The World of Islam*. Thames and Hudson, Ltd. 1992.

Mayr, Ernst. *The Growth of Biological Thought: Diversity, Evolution, and Inheritance.* Harvard University Press, 1982.

Popper, Karl. *Realism and the Aim of Science.* Edited by W. W. Bartley III. Routledge, 1983, 1992.

Wightman, P. D. *The Growth of Scientific Ideas.* Yale University Press, 1953.

## Notes

1. Both these scenarios are versions of a logical fallacy called *false cause*. In the first instance, one assumes that because two different events occur together, one causes the other. If you argued that men who drink milk for breakfast never get pregnant, you would be arguing from false cause. While a man's drinking milk at breakfast correlates with the absence of male pregnancy, no causal connection is possible. The second fallacy confuses a temporal relation with a causal one: It is called *post hoc, ergo propter hoc,* translated as "after the fact, therefore because of the fact." An additional instance of this fallacy follows: A man who was getting ready for bed in his stateroom on the Titanic turned off the light to sleep. Immediately afterward, the ship struck an iceberg and sank. "I should have left the light on," this survivor said, believing that his turning off the light bore a causal relationship to the ship's destruction.

2. Ernst Mayr, *The Growth of Biological Thought: Diversity, Evolution, and Inheritance* (Cambridge, Mass., London: The Belknap Press of Harvard University Press, 1982), pp. 30–32.

3. Norwood Russell Hanson, *Perception and Discovery: An Introduction to Scientific Inquiry* (San Francisco: Freeman, Cooper & Company, 1969), p. 158.

4. William P. D. Wightman, *The Growth of Scientific Ideas* (New Haven: Yale University Press, 1953), p. 220.

5. James Gleick, *Chaos: Making a New Science* (New York: Penguin Books, 1988), p. 8.

6. Thanks to Patricia Hartz, University of California, Irvine, for providing this systematic explanation of induction and deduction with copper as the example.

7. In the physical sciences, generalizations tend to be considered as "law." In the biological sciences, generalizations do not to extend to larger workings of the universe but are usually made with a more restricted application. The word *law* is not intended to suggest that this generalization is absolutely fixed forever and for all time or must be arbitrated in a formal hearing to dispute. *Law,* in this context, means a generalization.

8. See Gary Jason's *Introduction to Logic* (Boston, London: Jones and Bartlett Publishers, 1994), pp. 70–73.

9. To bring induction into the domain of a pressing social problem, one need only look to the AIDS epidemic for an instance of the danger created by mixing up high probability with proof. When the epidemic began, it was initially recognized in populations of homosexual men. Per the operation of induction, a generalization was made on the basis of discrete instances.

As such, it was considered to be a "gay disease," identified, until 1982, as GRID (gay related immune deficiency). Paula Treichler, in "An Epidemic of Signification," writes: "The question 'Why are all people with AIDS sexually active homosexual males?' might more appropriately have been '*Are* all people with AIDS sexually active homosexual males?'" She continues: "It is widely believed (not without evidence) that federal funding for AIDS research was long in coming because its chief victims were gay or otherwise socially undesirable" (*AIDS: Cultural Analysis, Cultural Activism*, edited by Douglas Crimp, [Cambridge, Mass., London: The MIT Press, 1988] p. 43, n. 22).

10. Mayr, *The Growth of Biological Thought*, p. 28.
11. Francisco J. Ayala, "The Criterion of Demarcation," in *Evolution*, eds., Theodosius Dobzhansky, Francisco J. Ayala, G. Ledyard Stebbins, James W. Valentine (San Francisco: W. H. Freeman and Company, 1977), p. 479. Previous versions of *Writing Science through Critical Thinking* omitted this bibliographical information.
12. Hanson, *Perception and Discovery*, p. 241.
13. Karl Popper, *Realism and the Aim of Science*, edited by W. W. Bartley III. From *Postscript to the Logic of Scientific Discovery* (London and New York: Routledge, 1983, 1992), p. xxxi.
14. This material derives from the manuscript, "An Interdisciplinary Guide to Critical Thinking," by my colleague at the University of California, Irvine, Dr. Patricia Hartz.
15. Galileo Galilei, "The Starry Messenger," in *Discoveries and Opinions of Galileo*, translated and introduced by Stillman Drake (New York: Doubleday, 1957), pp. 38–39.
16. Andrew Wear, introduction to William Harvey's *The Circulation of the Blood and Other Writings*, translated by Kenneth J. Franklin (London: J. M. Dent & Sons Ltd, 1990), p. xi.
17. William Harvey, "Movement of the Heart and Blood in Animals," in *The Circulation of the Blood and Other Writings*, p. 57.
18. Previously cited.
19. This material is adapted from Stephen Jay Gould's *The Flamingo's Smile* (New York: W. W. Norton, 1985), pp. 99–113 ("Adam's Navel").
20. I have provided a schema of a syllogism but not the body of thought that defines and limits the validity of any given syllogism. I refer you to other books or courses in logic for further elaboration.

# 2

## The Language of Science

*Science writing* can refer to two kinds of writing: writing *about* science (i.e., scientific subject matter) or writing *in the context* of science.

It is possible to write about science in a non-scientific way: Different written forms can take science as a subject about which something is said. An ode to the atom, a song celebrating the elements, a novelistic reconstruction of the discovery of X-rays—all of these could, strictly speaking, be considered types of writing that take science as their *subject matter*. The *form* of that subject matter would derive its structure and conventions from different genres like poetry, fiction, drama, and the essay. In other words, "science writing" could describe a type of writing in which the subject matter of science found a context within the conventions of literature.

But *science writing* may refer to another kind of writing, and it is this kind of writing that concerns us. Science writing may refer to writing that takes place within the context of science, governed by a specified set of conventions. Those conventions will be elaborated below. Because science writing refers to the manner of presentation, and not the topical content, any subject is a possible subject for scientific treatment.

Understood in this sense (writing generated within the context of a specific methodology), science writing belongs to category of writing called *technical writing* (as opposed, for instance, to the category of writing called *literary writing*). Technical writing refers to types of writing that report factual information in an objective manner, often for a targeted audience; it is a broader category than science writing. Science writing, as here defined, concerns itself with describing observations and/or results, and the manner by which they are derived through experimental or non-experimental methods, including classification, statistical analysis, mathematical analysis, and comparative studies. Such writing will now be stipulated *scientific writing* to differentiate it from types of writing that take science as a subject rather than as a context. Chapter 6 will provide a detailed discussion of the formal components of a scientific paper; this chapter will describe the attributes of technical communication.

## Technical Writing

Because technical writing presents factual information objectively, technical communication emphasizes the message. In this context, language functions as the means to an end, that of presenting information clearly and concisely. The writer does not encourage you to remark on the great beauty of the language; language should be transparent, a vehicle for the meaning. In fiction, poetry, and non-technical prose, the writer has a different end in mind. Sometimes these forms will communicate a subjective impression of an object or phenomenon; the writing may call attention to the language itself or the writer may give animate or human characteristics to non-human subjects. Sometimes non-technical prose is oriented to persuade the audience to take a particular point of view; in such cases, facts may be employed to solicit your approval. Such writing presents facts subjectively not objectively.

A casual glance at James Tate's "The Blue Booby" (below) tells you that it is not technical writing. In poetry, sentences or phrases are broken up and set on different lines. Line breaks are a poetic punctuation mark; they may direct your attention to the meaning, to the rhyme, to an image at the end of the line. The form signals you—through line breaks instead of full paragraphs—to read the words in a certain way. Line breaks are part of the code that says "This writing is poetry."

*"The Blue Booby"*[1]
   *James Tate*
The blue booby lives
on the bare rocks
of Galapagos
and fears nothing.
It is a simple life:
they live on fish,
and there are few predators.
Also, the males do not
make fools of themselves
chasing after the young
ladies. Rather,
they gather the blue
objects of the world
and construct from them

a nest—an occasional

Gaulois package,
a string of beads,
a piece of cloth, from
a sailor's suit. This
replaces the need for
dazzling plumage;
in fact, in the past
fifty million years
the male has grown considerably duller,
nor can he sing well.
The female, though

asks little of him—
the blue satisfies her
completely, has a
a magical effect
on her. When she returns
from her day of
gossip and shopping,
she sees he has found her
a new shred of blue foil:
for this she rewards him
with her dark body,
the stars turn slowly
in the blue foils beside them
like the eyes of a mild savior.

Even if you rearranged the text of the poem and changed the line breaks—

> The blue booby lives on the bare rocks of Galapagos and fears nothing. It
> is a simple life: they live on fish, and there are few predators. Also, the
> males do not make fools of themselves chasing after the young ladies.
> Rather, they gather the blue objects of the world and construct from them
> a nest—an occasional Gaulois package, a string of beads, a piece of cloth,
> from a sailor's suit. . . .

—the passage would remain non-technical.

It would no longer be poetry; it would be prose—non-technical prose
because the lines continue to convey the writer's *subjective* impression of
the subject matter. The behavior of the booby is represented in human
terms: Males "do not make fools of themselves chasing after young ladies"
(only a human could be a fool or a lady); the female gossips and shops. The
details used to describe the booby are selected and organized to enhance
the effect of this anthropomorphism. "The male has grown considerably
duller, nor can sing well"—these details are not the systematic part of a
description but are, instead, specifically chosen for an effect. We do not
receive other information about the evolutionary changes that have occurred in the past fifty million years; the writer has selected some information and excluded other information. The passage does not present us with

data but with anecdotal detail. So the description is not objective, nor are facts reported that can be verified.

In the following passage, taken from a technical work on boobies and gannets, the writer provides an objective account of courtship behavior.

> [A]n important context of Sky-pointing* is that of a male advertising his sexual receptivity to a female—overflying females commonly elicit it from males—[but] it has become equally or perhaps more important as a mutual display between partners that have already formed a bond. The term mutual is used to indicate that partners may actually display simultaneously—one like the mirror image of another—rather than merely one after the other, which I have called reciprocal. In the blue-foot both mutual and reciprocal Sky-pointing are strongly developed, the pair facing each other and repeatedly performing this bizarre display.
>
> A further aspect of Sky-pointing in a sexual context is its use in short-term co-ordination of sexual activity. Thus, it frequently precedes copulation and may be assumed to play a role in making the female receptive (or conversely, if initiated by the female, in stimulating the male).[2]

This passage provides an objective rendering of the courtship behavior of the blue booby. The writer describes what he has objectively observed and regularized in a pattern of courtship display. While Tate represents the behavior of the booby as a kind of domestic romance, Nelson portrays it as a pattern of behavior. Because no one knows what a booby thinks or feels, Tate's treatment animates them in human terms; he projects his own view of their interaction into the description.

Nelson departs from a purely objective approach only at one point: He remarks that the booby's sky-pointing behavior is "bizarre." *Bizarre* is a word that reflects his subjective opinion. Does the intrusion of subjective detail undercut the objectivity of a technical piece?

This instance alerts you to a qualification: the distinction "technical/non-technical" should be seen as a continuum with gradations rather than as absolute distinctions.

### Reporting Factual Information

Reporting factual information requires you to produce information that can be verified. To convey factual information, use words that are specific and concrete. Give precise measurements, dimensions, condition, speed, molarity, and so on.

---

*   Sky-pointing: a gesture assumed by the booby's extending its neck lengthwise and pointing its bill to the sky, accompanied by a forward rotation of the wings at the shoulder and a lateral extension of wings.

Be precise in your detail and give detail. Because the communication of information is the core of your writing task, your writing should be clear (your point is expressed directly) and concise (the most economical form of writing is used to convey your point).

## Objectivity and Tone

Scientific writing differs from non-technical writing in that the former is objective in tone. An objective tone is built up from words that are concrete, specific and precise. You can write about *any* subject as long as you present your discussion in an objective manner. There is no subject that cannot be treated technically. Love can be discussed in terms of the attractive qualities of pheromones; beauty may be discussed as a feature of patterned mathematic relations. In *The Astonishing Hypothesis: The Scientific Search for the Soul*, Francis Crick treats spiritual matters from the standpoint of scientific methodology. Modern (non-representational) art can be discussed objectively in terms of optics as a repudiation of linear perspective. The fact that anything *can* be written about objectively does not exclude the fact that it may also be written about subjectively. Francis Crick's treatment of the soul does not disallow discussion of the soul in reverential terms in other works; but a reverential attitude would disqualify such writing from scientific writing, just as a romantic attitude conveyed in writing about love or an awe-inspired treatment of beauty would exclude those writings from scientific writing.

Objectivity is conveyed through *tone*. *Tone* refers to the writer's attitude toward the subject, toward the audience, or toward himself. Tone is perceived through the writer's diction or word choice. An objective tone is created by words that have a clear and specific *denotation* (dictionary definition) and a neutral *connotation* (associations the word carries). For instance, if you were to refer to a horse as a "nag," *nag* would carry the sense that this horse was a broken down equine; it would carry a negative linguistic charge. *Steed* also refers to a horse, but this term carries a positive linguistic charge by the association of *steed* with the qualities of nobility, chivalry, and romance. The objective term, that is, the term with a neutral connotation, is the word *horse*.

Your writing also conveys your attitude to your audience. If you misjudge the audience level, you could come across as patronizing them. If you indulge in sarcasm or irony, you may convey the attitude that you do not take your audience seriously.

## Objectivity and the Passive Voice

Compare and contrast the phrasing in the following two sentences:

| Active Voice | ⇒ | I weighed the sample [4 words] |
| Passive Voice | ⇒ | The sample was weighed by me [6 words] |

In the active voice, the agent of the action takes the subject position in the sentence, and the receiver of the action takes the direct object position. Thus, in *I weighed the sample*, the do-er, "I," is the subject of the sentence. The object, i.e., the receiver of the action, is "sample." In the passive voice, the object of the action takes the subject position in the sentence; the agent of the action, "I" is noted as an object of the preposition or it may be omitted.

Objections to the passive voice note that it is wordy and vague. In cases where the agent is included in the sentence as the object of the preposition, the passive voice is wordy, in the strictest sense that the passive voice requires more words to present the same information as the active. In cases where the agent is deleted—"by me" is omitted—the passive receives the charge of being vague. The context of the writing defines the desirability of including the agent or not. In reporting the methods of an experiment, the focus is not on the person who conducted the experiment but on the experiment itself; the passive voice in such an instance is a better option than the active because the passive emphasizes the action not the agent.

Some writing scenarios are inappropriate for the passive voice; in other scenarios the passive may be most apt. In a lab report, the inclusion of the agent would be awkward and misleading in emphasis: "I measured the sample, then I weighed it, and then I heated it to 310 degrees." To express the same idea but to shift the accent, it is appropriate and desirable to use the passive voice, as in the sentence: "The sample was weighed, measured, and heated to 310 degrees." The "I," while not present in the sentence, is implied.

**Persona**

While the transmission of information forms the primary purpose of your writing, a secondary effect is produced. Your tone creates a picture of you, the writer. This picture is called a *persona*. Ideally, your persona should convey personal and/or professional integrity. The writer's integrity inspires the reader's confidence in your work and makes your research credible. The writer's trustworthy persona can be established by any one or combination of the following: (1) *authority* (you've read the books, you've done your homework); (2) *position* (professional status); (3) *education* (you have a degree in your subject); (4) *experience* (you've done this before so you know what you're talking about); (5) *reputation* (everyone knows you can be trusted).

Note: the persona is not a conscious construction. It is an *effect* of your writing, conveyed by and through the discussion of your material. Beginning student writers sometimes view the task of presenting a scientific paper with some trepidation, especially if that paper is presented to an expert in the field. Beginning writers can establish their authority through the literature search. A conscientious search and thoughtful consideration of other research in the field will be reflected in the scientific paper and, in turn, will reinforce trust between the writer and the reader. In this creation of trust, the writer establishes his or her authority.

Your audience equates you with your writing. If your literature search has been thorough, your experimental design appropriate, your discussion relevant, and your interpretation sound, then you create the grounds of your own authority for the reader. Conversely, it is possible to subvert that authority: if your writing is sloppy, your organization chaotic, your paper covered with coffee stains, and your punctuation random, you create the image of a person who is careless. It would not be unreasonable for your audience to infer that a person who does sloppy writing does sloppy science. Take care with both your writing and with the physical appearance of your presentation. Avoid sarcasm or humor; self-irony tells your audience that you do not take your writing seriously. Use the spell-check and the grammar-check on your computer. Spelling errors in a formal paper are one of the most glaring signs of carelessness because they are so easy to correct.

### Analogy versus Metaphor

Analogies may be used in technical writing to explain or elucidate a point in the writing. Written for the purpose of clarification, an analogy makes a comparison between things or events. The statement "Spider silk is elastic like rubber" asserts a likeness between an organic substance, spider silk, and a substance well known for its elasticity, rubber. The analogy renders a point by stressing the likeness between elements in the comparison.

In contrast, a metaphor or simile makes a comparison that defamiliarizes, or makes strange, a conventional understanding of the terms linked together through it; in this regard, it calls attention to itself and may be used to evoke a response or express an emotion. For example, in Sonnet 24, Shakespeare compares the seeing function of the eye with the rendering capability of the artist: "Mine eye hath played the painter and hath stelled / Thy beauty's form in the table of my heart." The eye in this line is compared to the painter; the act of seeing has drawn (stelled) the beloved's face on the notebook (table) of the heart. Metaphor here underscores the indelible impression left by the beloved by the ascription of agency to the passive operation of sight.

The purpose of analogy in science writing is clarification. Clarification is rendered by the parallel drawn between some thing that is not well known and some thing that is well known. How elastic *is* spider silk? You know the answer: It is as elastic as rubber.

Analogies may be simple comparisons, one phrase or one sentence long, or they can extended. In the following passage, George Gamov, in *A Biography of Physics*, compares the change in electron orbital with a gear shift in a car:

> If radiant energy can exist only in certain minimum amounts or the multiples therefore, why not make the same assumption concerning the mechanical energy of electrons circling the nucleus? In this case, the motion of electrons in the normal state of an atom would correspond to a larger number of these mechanical energy quanta. Thus, *an atomic mechanism should behave, in a way, as an automobile transmission box:* one can put it in low, in second, or in top gear, but not in between.[3]

In scientific writing, an analogy does not stand out or call attention to itself; instead, it clarifies or explains. If you wanted to be hard-minded about the relationship between metaphors and analogies, the truth is that they are very much alike structurally but differ greatly in function: analogies make familiar, metaphors make strange.

# Terminology

Technical writing reports factual information. To convey facts correctly and to demonstrate objectivity, you will often have to explain your reasoning. To this end, you need a clear understanding of the difference between a fact, an inference, and a judgment.

### Fact

A fact is a statement thought to be true. It is verifiable. It does not change, although it may be subject to interpretation. Its adequacy obtains here and at the South Pole. The speed of light is constant both here and at the other side of the universe. A ball dropped from the Leaning Tower of Pisa and a ball dropped from the temple at Teotihuican will always fall *down*.

There is one qualification, however, to this definition of a fact. In the history of science, some facts which were thought to be true were later discovered to be erroneous. For instance, the Ptolemaic universe (a classical paradigm) placed the earth at the center of the solar system and described the sun and other planets as orbs that circled the earth. Even though we know that the sun stands at the center of the solar system, those earlier navigators could still arrive at their planned destination by using information that was not based on a "true" fact.

Those navigators had a functional model of the universe, which, though not true, was adequately descriptive. For the purpose of a working definition, however, we will consider a fact to be a point of information that all people will *agree* is true, i.e., a fact has been demonstrated to most adequately describe events in the world. As discussed in Chapter 1, truth is stipulated in the sense of a provisional truth and is not to be conceived as Truth, a quasi-metaphysical absolute term on which much faith is staked.

## Inference

An inference is what we make of the facts. An inference is a reasoned interpretation. Because it is reasoned, someone who replicated your method of reasoning should logically arrive at the same conclusion. Consider, for instance, the relations proposed in the following sentences:

(1) To pass the class, you must pass the final.
(2) Joe failed the final.
(3) Therefore, Joe will not pass the class.

The conclusion (3) that Joe will not pass the class does not reflect a bias or a personal judgment on this person. Rather, it is the logical conclusion of a series of steps and conditions.

Consider the following logical argument:

(1) $a=b$
(2) $b=c$
(3) Therefore, $a=c$

This is a logical pattern called a syllogism. Anyone who performed the logical steps in the same order would arrive at the same conclusion. The major premise (1) and the minor premise (2) are clearly specified, from which the logical and necessary result must be (3).

## Opinion and Judgment

An opinion is a view you hold based on personal preference. A judgment is an assessment of value based on a personal opinion. For instance, the statements "Blueberry pie is my favorite" (opinion) and "A BMW convertible is the best car in the universe" (judgment) reflect personal preference. As expressed here, these statements cannot be proved objectively. Although they may be disputed, there is no binding necessity that can make blueberry pie *your* favorite if you already hate it.

You may be tempted to argue that you could prove a BMW was the best car in the universe. Shun this temptation. You would have to rephrase this statement to include the criteria of evaluation. The troubled term in that statement is *best*. Best could be judged in terms of most efficient design, cost, performance, or other criteria, but because the statement does not specify objective criteria, that statement remains a judgment.

## Exercise

Answer each of the following and write a response in a few objective words.

(1) Describe a piece of chalk.
(2) Given that a=b and that b=c, does a=c?
(3) What do you think of North American environmental policy?
(4) What do you think of modern art?

(1) We will agree that chalk is white, or flaky, dusty, or made of calcium. Normally, instructions for a description elicit a statement about the essential attributes. These attributes of substance are facts because they provide observable or measurable data. They are verifiable. Anyone else who was asked the same question would have the same answer or would provide an answer consistent within a range of possibilities (after all, only a few words were specified in your description). To some extent, facts are a product of general agreement. Everyone agrees that the facts about chalk are true. If I said that chalk was a symbol of beauty, you might not agree with me. In fact, if I insisted that it was a *fact* that chalk was a symbol of beauty, you would question my ability to understand facts; the object would not be suspect; the person would be. A description of chalk that specified its physical attributes would make an *empirically true* statement. The chalk can be measured (length) or analyzed (calcium carbonate).

(2) The statement that a=c is *logically consistent*. Anyone who performed the same series of steps would arrive at the same conclusion. Inferences are logically verifiable. Notice that this example specifies a starting premise.

(3) Any decision on this question would depend upon your starting premise. This question elicits a *subjective interpretation*. While you can follow such an interpretation, no logical necessity can force your agreement with it. Your starting premise is a matter of personal choice.

(4) Your view of modern art is subject to your personal opinion. A personal opinion need have no basis in fact nor require any justification. You do not need reasons to have an opinion. Although you might have reasons to account for your like or dislike (It is pretty. It is ugly.), those reasons are ultimately subsumed to matters of personal taste. It is possible to have an opinion about any subject but you cannot prove your opinion or convince others to have the same opinion through reasoning.

## True and Correct Language

When we say that a sentence is written correctly, *correct* refers to a context of language rules. A sentence is written in accord with the grammar and the punctuation that make the meaning intelligible. *Correct* means that language conforms to the rules of grammar, diction, and punctuation. Consider the following examples:

(1) A red onion sang pearls against a light green moon.
(2) A a against green light moon onion pearls red sang

What is the difference between (1) and (2)?
They both use the same words.
Example (1) is a *correct* sentence.

It observes the rules for syntax. It has a subject, a verb, and a direct object. Adjectives stand in relation to the words they modify. Example (2), while using the same words, arranges them in alphabetical order. You could recognize that (2) is not a sentence. The words must be organized in a certain way for us to understand they are to be read as a sentence.

- The way words (and sentences) are put together gives us indirect instructions about how to read them.

Word order gives directions or evokes a code: *read this as a sentence.*

The words alone provide us with little meaning. But their arrangement, in (1), instructs us to read them as a meaningful unit, a sentence. Similarly, earlier in this chapter, the line breaks in the poem gave the instructions: *read this as a poem.*

While *correct language* obeys laws of syntax, grammar and punctuation, *true* language is different. When we say a statement is *true*, we mean that a description or statement conforms to the world as we know it or as it is generally perceived to be. Technical language creates a model that stands in correspondence to the world. In contrast, consider the following sentence:

(3) Flying pigs sing orchids to sleep on submarines.

This sentence is grammatically *correct*. It has a subject, verb, and object that are arranged in a way that makes a sentence. But it is not a *true* statement because pigs neither fly nor sing. While orchids may suffer a slower metabolism at night, we do not generally call such a state sleep. This sentence makes a kind of poetic sense; but when you compare its information with the general knowledge you have about animals and plants, it does not make a *true* statement. Within the limits of the world as we know it, pigs do not possess the capacity for flight or vocalization. So this sentence, while *correct* (adequate to the rules of grammar) is *not true*.

The language of science must be both *correct* and *true*. It must possess both conformity to the rules of grammar, punctuation, and spelling, and it must make a statement that is adequate to events in the world.

## Limitations on True Language

While scientists strives to use language that is adequate to our knowledge of the world (*true*), it is possible for language to be false to reality and yet describe an aspect of the world. For instance, you would probably understand the statement, "The sun came up this morning," even though it is literally false. Because the earth rotates, the sun only appears to move. It is possible to make a statement that everyone understands but which is not precisely descriptive of the events as they are.

While language can *report* a fact, it does not *equal* a fact. We attempt to build a model of the world as we perceive it or know it, but language ultimately remains a model, susceptible to the capabilities and restrictions of other models. For example, while we conveniently attribute color as a property of substances—The sky is blue. Roses are red. Clouds are white.— color actually describes a select spectrum of light waves as they are decoded by the optic nerve and registered on our brain. Does the color exist in light? Does it exist in our brain? Does it exist in the thing itself? Does color exist in a dark room?

Before you spend too much time pondering these questions, bear in mind the larger point: Language gives us a model that *attempts* to be adequate to the world but which *may not* be adequate to the world. Does this mean that people who try to use language at all are trying to cope with an impossible task? Are we fools for thinking we can use language to communicate anything? Such a cynical approach is not accurate either. We can communicate with each other. Remember, language is a social tool or instrument designed to help us interact with other people. Consequently, we can all *agree* that meaning and meanings exist. If I say "dog," you will know, because you have been taught, what *dog* is or means. If I say a foot is twelve inches long, you can comprehend the dimensions because we have all learned common conventions and common words. So while language may not always be adequate to the world, it is still possible to communicate with each other. Language use is conventional and the stipulation given above about true language is one such convention.

## Two Criteria for Science Writing

Science writing must satisfy two criteria: it must be *empirically true* and *logically consistent*.

Let us go back to the syllogism, denoted by the scheme: (1) a=b (2) b=c (3) a=c. If we substitute sentences for letters, we can produce the syllogism you have previously seen in Chapter 1:

All copper conducts electricity.
This piece is copper.
>    Therefore, this piece conducts electricity.

This order of sentences reflects steps in thought, and those steps will, per the rules of logic, consistently produce a common conclusion.

But logical consistency is not the only criteria for science writing. A statement may be logically consistent but not empirically true.

C. L. Dodgson was a British mathematician who wrote two books on logic. (Dodgson is better known as Lewis Carroll, the author of *Alice in Wonderland*.) The following syllogism is adapted from his book, *Symbolic Logic*.

All chickens speak French.
This creature is a chicken.
> Therefore, this creature speaks French.

What is wrong with this conclusion?

This syllogism is logically consistent, but it does not arrive at a conclusion that is accurately descriptive of the world as we know it. *This creature speaks French* is the logical conclusion (and a grammatically *correct* statement), but we know that chickens do not speak French or English or Spanish. Chickens do not speak at all. Therefore, the first premise and the conclusion fail to conform to our knowledge of the world. While this syllogism makes a logical inference, it is not adequate to the facts we have about animal behavior.

If you accept the premise, the starting statement, logic compels you to accept the conclusion.

But you do not have to accept the premise.

This syllogism starts with a premise that fails to correlate with facts. This premise is not a remark that derives from a fact; it derives from a sentence that is grammatically correct but logically nonsense.

Both these syllogisms are logically consistent in their train of reasoning; however, each builds upon different kinds of premises. The premises in the copper syllogism are true statements about the world; in contrast, Dodgson's syllogism proceeds from a nonsense premise.

Scientific reasoning rests on premises based on facts and proceeds logically. Such reasoning is therefore both empirically true and logically verifiable; arguments based on such reasoning are said to be sound. Both criteria must be met. Logical consistency alone is insufficient.

You ensure that your writing is empirically true when you include data, measurements, quantities, and conditions; these specifications are contained in the Materials and Methods section. It is also important that your paper include information about the instruments used to make those measurements. You ensure logical consistency by explaining your results in a reasoned manner.

## Exercises

### 1. Technicalities

Go back through this chapter, and, after synthesizing what you read, make a table to compare technical writing with non-technical writing on the following points: purpose, tone, subject matter, style, or any other points you may note.

## 2. Objective or Not?

What is the subject matter? How is the subject matter treated? Recall that the distinguishing mark of scientific writing lies in the treatment (the "how") and not in the subject matter (the "what"). (A description of an object does not make a description objective.) Using the distinction between subject matter and subject treatment, comment on the subjectivity/objectivity of passages below. Refer to particular words or phrases in the text to support your view. Then rewrite the subjective accounts to make them objective.

(a) At tea, however, a shape began to emerge which brought back our spirits. Three chains twisted around each other in a way that gave rise to a crystallographic repeat every 28 Å along the helical axis. This was a feature demanded by Maurice's and Rosy's pictures, so Francis was visibly assured as he stepped back from the lab bench and surveyed the afternoon's efforts. Admittedly, a few of the atomic contacts were still too close for comfort, but after all, the fiddling had just begun. With a few hours' more work, a presentable model should be on display.[4]

(b) When the NMDA glutamate channel opens, it allows the passage of not only sodium and potassium atoms but also an appreciable amount of calcium ions (Ca++). These incoming calcium atoms appear to be the signal that initiates complex chains of chemical reactions that at the moment are only partially understood. We can now begin to see the beginnings of the explanation of cognitive processes, such as memory, in terms of molecular events.[5]

(c) He was a very small frog with wide, dull eyes. And just as I looked at him, he slowly crumpled and began to sag. The spirit vanished from his eyes as if snuffed. His skin emptied and drooped; his very skull seemed to collapse and settle like a kicked tent. He was shrinking before my eyes like a deflating football. I watched the taut, glistening skin on his shoulders ruck, and rumple, and fall. Soon, part of his skin, formless as a pricked balloon, lay in floating folds like bright scum on top of the water: it was a monstrous and terrifying thing. I gaped bewildered, appalled. An oval shadow hung in the water behind the drained frog: then the shadow glided away. The frog skin bag started to sink. I had read about the giant water bug, but never seen one. "Giant water bug" is really the name of the creature, which is an enormous, heavy-bodied brown beetle.[6]

(d) Hemipteran mouthparts are modified to form a piercing beak. Whereas most terrestrial bugs are phytophagus, using the beak to pierce plant tissue, aquatic hemipterans (including many *Corixidae*) inject enzymes to liquefy the tissues of animal prey that are sucked up through a food channel in the beak. Prey normally consists of small insects and crustaceans, although there are accounts of giant water bugs (Belostomatidae: Lethocerinae) attacking and subduing frogs, fish, and water snakes. The largest aquatic insects are members of the genus *Lethocerus*.[7]

## 3. Analysis of Ethos

*Ethos* refers to the individual's character. Ethos is conveyed through the writer's *persona* (image) created through writing.

We have said before that any subject can be treated from a technical point of view, that no subject matter is intrinsically forbidden. This assertion appears straightforward, and we have come to believe that we live in a such a modern time and place that any subject can be discussed with objectivity; yet, there are still topics that might appear on the margins of legitimate scientific inquiry.

In the early 1980s, the Dean of the Princeton School of Engineering/Applied Science was asked to write a paper on ESP. A highly regarded professional journal, the *Proceedings of the Institute of Electrical and Electronics Engineering (IEEE)*, solicited this article from him. The Dean of the Princeton School of Engineering did write that paper, with surprising results.

But before we consider what Dean Jahn wrote, let us spend a few minutes considering his position. R. G. Jahn was the dean of a hard-science discipline, Applied Science. He was also the dean at a prestigious university, Princeton. His official position made him a figure with much authority. He was invited to report on a subject that was probably seen as a marginal subject, perhaps something like magic or voodoo. Dean Jahn could have refused, but he did not. He produced a comprehensive, analytic, documented essay on the subject.

In the following sections, you will see three instances of introductory writing: the introduction provided by the journal, the abstract, and Jahn's introduction to the paper are all designed as types of writing that acquaint the reader with the subject. Read through the passages and then answer the following questions.

a. Look at all of the passages. According to each of them, what has been the previous professional attitude toward ESP research?

b. What has been the traditional popular view? Does the popular view conflict or coincide with the professional view?

c. Compare and contrast the tone in the journal's introduction with that in Jahn's introduction.

d. How did Jahn get involved in the project? How did his attitude change from the time he became involved in this subject to the time that this paper appeared?

e. Does Jahn currently spend all of his time now on ESP? How deep is his involvement? Why does he explain this?

f. Aristotle once said, "A speaker worthy of belief is the most potent of all the means of persuasion." On the basis of c. above, how would you characterize Jahn's ethos? Is he "worthy of belief"? What aspects of his writing can you point to in support of your view? Does Jahn come across as an eccentric administrator? As a careful scientist? Do you feel that he sacrificed any of his professional authority in dealing with this subject?

## I. *IEEE* Introduction to R. G. Jahn, "The Persistent Paradox of Psychic Phenomenon."

The introduction to this essay, which appears in the introduction of Vol. 70, No. 2, *IEEE*, February 1982, describes Jahn's contribution as follows:

**The Persistent Paradox of Psychic Phenomenon: An Engineering Perspective** *(Invited Paper), Jahn,* page 136—The persistent fragmentary indication of the possible existence of psychic phenomena requires us, in the interests of scientific integrity and intellectual honesty, to apply the same spirit of objective inquiry that we have focused on the more traditional and accepted branches of science and technology.

It was nearly six years ago—in March 1976, to be exact—that the *Proceedings* took the unusual step of publishing a paper on the psychic phenomenon called remote viewing. That paper was received with great interest, considerable skepticism, some hostility, and even a bit of shock. In the intervening years, no breakthroughs have occurred to prove or disprove the existence of this and other psychic phenomena, but interest and research have continued unabated. Where does the research on these topics stand now? It was with this question in mind that the paper in this issue was invited.

This paper reviews the history, nomenclature, conceptual organization, and status of the generic field of psychic phenomenon; presents a few detailed examples of contemporary research that could have ultimate technological implications; and attempts to offer a balanced view of the viability and value of continued study of this fascinating and controversial field.

This paper is believed to represent the first attempt at a comprehensive engineering survey of the field. Most readers will likely approach it with a healthy blend of curiosity and skepticism. It is hoped all will find it informative and thought provoking.

## II. R. G. Jahn's Abstract

Although a variety of so-called psychic phenomena have attracted man's attention throughout recorded history, organized scholarly effort to comprehend such effect is just one century old, and systematic academic research roughly half that age. Over recent years, a sizeable spectrum of evidence has been brought forth from reputable laboratories in several disciplines to suggest that at times human consciousness can acquire information inaccessible by any known physical mechanism (ESP), and can influence the behavior of physical systems or processes (PK), but even the most rigorous and sophisticated of these studies display a characteristic dilemma: The experimental results are rarely replicable in the strict scientific sense, but the anomalous yields are well beyond chance expectations and a number of common features thread through the broad range of reported effects. Various attempts at theoretical modeling have so far shown little functional value in explicating experimental results, but have served to stimulate fundamental re-examination of the role of consciousness in the determination of physical reality. Further careful study of this formidable field seems justified, but only within the context of very well conceived and technically impeccable experiments of large data-base capability, with disciplined attention to the pertinent aesthetic factors, and with more constructive involvement of the critical community.

## III. Introduction: R. G. Jahn, "The Persistent Paradox of Psychic Phenomenon"

In the introduction, Jahn explains the history of his involvement in the project. This segment comes from his formal paper.

> . . . I venture to begin the most extraordinary writing task I have yet attempted: to respond to the request of the Editors of this journal for a critical review of the status and prognosis of scientific research into so-called psychic phenomenon. I do so with some trepidation, first because the topic is far from my principal line of scholarship and my involvement with it has been brief and tightly circumscribed, and second, because of the intensity of reactions any commentary on this subject tends to call forth from many quarters.
>
> For these reasons, it may be well at the outset to specify my perspective on the field and the purpose that I hope this article will serve. My formal training is that of an engineer and applied physicist, and the bulk of my research has concerned a

sequence of topics in the broad domain of the aerospace sciences: Fluid mechanics, ionized gases, plasma dynamics, and electric propulsion. In my present position as Dean of the School of Engineering and Applied Science of Princeton University, I have occasion to be involved with an even broader selection of topics selected for undergraduate independent projects, and it was in that context some four years ago that I was requested by one of our very best students to supervise a study of psychic phenomena. More specifically, this young lady proposed to bring her talents and background in electrical engineering and computer science to bear on some experiments in controlled low-level psychokinesis. Although I had no previous experience, professional or personal, with this subject, for a variety of pedagogical reasons, I agreed, and together we mapped a tentative scholarly path, involving a literature search, visits to appropriate laboratories and professional meetings, and the design, construction, and operation of simple experiments. My initial oversight role in this project led to a degree of personal involvement with it, and to a growing intellectual bemusement to the extent that by the time the student graduated, I was persuaded that this was a legitimate field for a high technologist to study and that I would enjoy continuing to do so. I have since assembled a small professional staff, secured the requisite funding from a few private sources, and undertaken a modest experimental program in selected aspects of the field that could ultimately have some engineering implications. I should emphasize that my fractional involvement with this program remains quite minor in comparison to my other responsibilities, and that the work is still very preliminary and tentative, but it provides the base of cognizance for my broader observations on the field as a whole.

The intention of this article is to provide some balanced perspective on the modern status of this conceptually and logistically difficult subject. Certainly no field of scholarly endeavor has proven more frustrating, nor has been more abused and misunderstood, than the study of psychic phenomena. Dealing as it does as much with impressionistic and aesthetic evidence as with analytic substance, and carrying by its nature strongly subjective and numenistic overtones, it has been incessantly prostituted by charlatans, lunatics, and sensationalists, categorically rejected by most of the scientific establishment, and widely misunderstood by the public at large. Interspersed with this, and greatly encumbered by it, a pattern of legitimate effort to comprehend and utilize the purported phenomena has evolved to a point where some dispassionate assessment of its accomplishments can be attempted. (136–37).

## 4. Analysis of Graphic Illustration

George Stubbs was born (1724–1806) in Liverpool. His father was a currier, or leather dresser, and George must have been accustomed to tan yards and slaughterhouses at an early age. By the time he was 8, Stubbs was drawing with an amazing talent. He continued to make drawings of animals all his life. Although he tried to make a living by painting portraits and landscapes, his greatest success came from his publication of etched anatomical drawings. After traveling abroad, he eventually established himself in London where he began his work on *The Anatomy of the Horse* (published March 4, 1766). His work included plans for a series of drawings that compared the anatomy of an ape, a tiger, and a horse.

The following illustrations come from *The Anatomy of the Horse*. Notice that the line drawing of the skeleton contains letters and numbers; these are keyed to a

Figure 1 (overleaf spread): Second Skeleton Table (plates 88 and 89) from Stubbs's *Anatomy of the Horse*.

description of the labelled parts that occurs in the written text accompanying the illustration.

Formulate the criteria for technical writing and then apply them to Stubbs's etchings in order to make a determination about the technical or non-technical nature of these illustrations. Can you define criteria to distinguish an "objective" drawing from a "subjective" drawing?

## References

Crick, Francis. *The Astonishing Hypothesis: The Scientific Search for the Soul*. Charles Scribner's Sons, 1994.

Dodgson, Charles. *Symbolic Logic*. Macmillan, 1896.

Gamow, George. *The Biography of Physics*. Harper & Row, 1961.

Dillard, Annie. *Pilgrim at Tinker Creek*. Harper's Magazine Press, 1974.

Jahn, R. G. "The Persistent Paradox of Psychic Phenomenon: An Engineering Perspective." *IEEE* 70, No. 2 (1982).

Lannon, J. *Technical Writing*. Little, Brown, and Co., 1979.

Nelson, J. Bryan. *The Sulidae: Gannets and Boobies*. Oxford University Press, 1978.

Stubbs, George. *The Anatomy of the Horse*. Original Text and Modern Paraphrase by J. C. McCunn and C. W. Ottaway. Dover Publications, 1976.

Tate, James. *The Oblivion Ha-Ha: Sixty Poems by James Tate*. Little, Brown, and Company, in association with the Atlantic Monthly Press: 1967, 1968, 1969, 1970.

Ward, J. V. *Aquatic Insect Ecology: 1. Biology and Habitat*. John Wiley & Sons, Inc., 1992.

Watson, James D. *The Double Helix: A Personal Account of the Discovery of the Structure of DNA*. W. W. Norton & Co., 1980.

## Notes

1. James Tate, "The Blue Booby," in *The Oblivion Ha-Ha: Sixty Poems by James Tate* (Boston: Little, Brown & Co., in association with the Atlantic Monthly Press: 1967, 1968, 1969, 1970), pp. 5–6.

2. J. Bryan Nelson, *The Sulidae: Gannets and Boobies* (New York: Oxford University press, 1978), p. 555.

3. George Gamow, *The Biography of Physics* (New York: Harper & Row, 1961), p. 239. The emphasis is mine.

4. James D. Watson, *The Double Helix: A Personal Account of the Discovery of the Structure of DNA*, Norton Critical Edition, edited by Gunther S. Stent (New York: W. W. Norton & Co., 1980), p. 56.

5. Francis Crick, *The Astonishing Hypothesis: The Scientific Search for the Soul* (New York: Charles Scribner's Sons, 1994), p. 101.

6. Annie Dillard, *Pilgrim at Tinker Creek* (New York: Harper's Magazine Press, 1974), pp. 5–6.

7.  J. V. Ward, *Aquatic Insect Ecology: 1. Biology and Habitat* (New York: John
    Wiley & Sons, Inc., 1992), p. 11.

# 3

⌁

# Audience and Purpose: Developing a Perspective

The Instrumentation and Labor Improvement Plan (ILM) is part of the National Science Foundation's effort to strengthen undergraduate education by offering grants for the purchase of equipment. The NSF typically issues guidelines that specify the content and the format of the grant proposal. Among the many specifications was the following guideline for the summary section, that section used by the researcher to explain what would be done with equipment were the proposal funded.

> The Summary of Proposed Work should be a concise description of the project (not the proposal), limited to 22 single spaced lines of 12 point type. The summary should briefly tell the aim of the project, the major instruments which will be purchased, in what applications they will be used, and why the project is significant. Considerable care should be taken when writing the summary. The Summary is the reviewers' first impression of the project's merit. If the project is supported, the Summary will be published by the Foundation to inform the general public about its programs. Accordingly, *it should be written so that a scientifically literate lay person could understand the use of Federal funds in support of this project.*—NSF, Public Document, Program Guidelines, 1992 (my emphasis).

The content of the summary is explicitly delineated: "the aim of the project, the major instruments which will be purchased, in what applications they will be used, and why the project is significant." A list could be generated to respond to these specifications:

(1) The aim is . . .
(2) The instruments are . . .
(3) The instruments will be used for . . .
(4) The project is significant because . . .

But these guidelines specify more than content or the subject matter or the proposal; they also supply the context in which the request will be regarded. The general public will be a third party to this funding request. A specific level of language is indicated for that audience—scientifically literate lay (unspecialized) audience—but further, the purpose states that this audience should be able to *understand* the allocation of taxpayers' monies. While the *form* of the summary could be satisfied with a list, the *writing task* requires you to anticipate and address the response of the targeted audience in light of the purpose: their understanding. The targeted audience and your purpose in writing to them are interconnected in shaping the organization and presentation of your writing.

## Audience

Who will read your writing?

Is your scientific paper directed to someone who possesses your level of expertise? Are you writing because someone will have to make a financial decision on the basis of your request? Who will repair instrumentation you describe? Can you assume that a college-level audience knows calculus and thus equations are appropriate? Is the reader familiar with complicated graphics or would a pie charts be more appropriate? Will the audience be your intellectual peers?

*Identify the audience for everything you write.* You need not visualize a complete individual—hair color, eye color, height, weight. The purpose of identifying the audience is to key into a common level of expertise in order to work from a common language. Identify a typical audience, one which represents the generalized responses of a class rather than those of an idiosyncratic individual.

The targeted audience governs your approach. As a writer, you must bridge the information gap between what your audience already knows and what they need to know to understand your point. In the absence of a common vocabulary between the writer and the reader, the obligation lies with the writer to generate that language. This common language will be slanted for the reader's comprehension.

*Because you are the writer, the burden of intelligibility rests with you.* Write in the language your audience understands. When in doubt about audience's expertise, explain a bit but make your explanation succinct. Fill in the information gap. Never expect the audience to bone-up on their reading to be able to understand your work. Jargon is specialized language familiar only to experts; a general audience cannot be assumed to understand jargon.

Private codes are generally inaccessible; some poets or fiction writers have written in a language that comes from the author's personal code of language. This kind of artistic writing emphasizes the medium (or the writer's expressing emotions, feelings, personal thoughts) rather than the message; your goal is the opposite: emphasize the message. Your goal is to communicate your purpose to your specified audience.

Writing that misses the mark is writing that misjudges the audience.

## Purpose: What Do You Tell?

Writers write for a purpose; science writers have information, facts, knowledge, and interpretations to share. As a writer, you should be clear about your own purpose in writing. What do you want to tell?

In telling about a subject, tell about the subject at hand rather than about yourself. You may have encountered some personalities who leave you with a stronger impression about their personality than about the topic of discussion: "I didn't understand a word, but every word was brilliant" or "I understood everything but I don't believe a word." The information relayed must pertain to the topic under consideration. The writer remains unobtrusive—unobtrusive but authoritative (not authoritarian)—and in so doing preserves the objective tone discussed in Chapter 2. As you work through the thinking process of your research, question your own views. Establish that they are reasonable in order to inspire the reader's confidence in you as the basis for trust.

As you conceptualize the expertise of the audience, imagine what the reader could do with the information or data you report. How can the reader make use of the data? Will your audience be made cognizant of information about which they were formerly ignorant? Do you wish your audience to take action, ultimately, to follow your recommendations? Do you want the reader to be able to *do* something with the information you provided?

If writing a description of an object, should the reader be able to form a clear mental picture of the object you have described? Should the reader be able to use the description as a guide to operating the object? Will the reader be able to replicate your experimental procedure and your results? Are the specific results less important to the reader than the implications of those results? Are the results more important than the implications of the results because your reader is a colleague who will, by virtue of a shared context, understand the significance of your results. There is a difference between "I understand what you are saying" and "I understand what you are saying and I agree with you." You may be relaying information because you want to change that person's mind away from preconceived notions he or she already possesses on the subject. A number of ends may intersect.

While it is possible to anticipate a normative response to a piece of writing, the writer wields no absolute control over the reception to his or her paper. Do not mislead your reader by reporting inaccurate information for the sake of securing a desired response. Not only does false reporting violate the convention between writer and reader (your reader approaches your paper with good faith), but you risk destroying your credibility and that of your institution. The U.S. Department of Health and Human Services defines scientific misconduct as "fabrication, falsification, plagiarism, or other practices that seriously deviate from those which are commonly accepted within the scientific community. It does not include honest error."[1] Because writing invokes an implied contract between the writer and the audience, conventions and codes govern the interaction between you and your audience. The most fundamental convention between the writer and the audience is that of trust. Because scientific research is practically and intellectually collaborative, maintaining trust is no light matter.

## Audience and Purpose

While working in a laboratory, you come across a new variety of virus. After repeated trials and extensive documentation of your finding, you must decide how you will go about making your results available to others.

Dr. Perez, who runs the lab where you work, is facing a staff reduction and budget cuts; you do not know how you stand in her list of priorities, and your term of research in her lab is drawing to a close. Because you want to continue your research, and therefore wish to continue using the lab, you will have to request an extension of your privileges. Your request for an extension will be accompanied by an explanation based on the significance of your research. Dr. Perez will make a decision on your request.

Meanwhile, your best friend from high school also does related research in this field. The last time you met over the holidays, you discussed your different interests and became aware of the intersection of your lines of work. You know that your finding will benefit your friend's work just as her work has enhanced your own. She is a social peer as well as a professional colleague. When you go home for the weekend, your younger brother, who is interested in science, asks you about your research. He sees you as a role model and aspires to follow in your footsteps one day.

You are, in short, faced with different communication scenarios. You need to communicate certain facts of your finding, but the audience and your purpose differ in each case. Dr. Perez stands in a position of authority with respect to your work; you are accountable to her. She must assess the priority of your request in the context of others who also see their work as important to them as yours is to you. How do you fit into her lab? and how do stand in relation to others who have the same space/equipment needs that you do? Your position in this scenario is analogous to that of a researcher writing a grant proposal for a continuance to be read and judged by a review panel of experts. Your purpose in communicating with Dr. Perez is to define (1) what you need from her and (2) why your work merits further support.

Your friend is socially and intellectually an equal; a theoretical background and interest in experimental design link your fields. You both stand to benefit from the exchange of mutually informative discussion. Your colleague will understand the implications of your results at once. In this scenario, your colleague is analogous to that of a peer-reviewer, an expert in the field who can assess the scientific merit of your work, both on its strengths and in comparison with the merits of other related work in the field; the action you wish your colleague to take is that of critique and intellectual exchange. While your younger brother shares your interest, he does not possess the same level of expertise, a specialized vocabulary, or an understanding of the scientific concepts on which your work rests. You are the authority with respect to him. While the particular details of your work will be largely inaccessible to him, he will understand a general explanation of what you are doing and what it means, and thus you do not expect him to critique your work or to award funds for the continuation of your research.

HOW you communicate relates to WHAT you communicate relates WHOM you communicate to

Note: If you do not know who your audience is, assume it to be composed of your peers: those who possess the same expertise and language skills that you do. Your peers are not necessarily the other students in the same virology class on the tenth lecture. While you share the same physical conditions, or possess other commonalities, for the purpose of the paper, peers are intellectual equals.

## Kinds of Audiences

### Audience Classified by Levels of Expertise

For purposes clarification, we will consider audiences in five different categories.* Five is not an absolute number tied to a fact of nature; it simply offers the convenience of classification. If you rounded up, in general, the kinds of people who might be reading scientific reports, you could, as you wished, divide them into five categories based on the use each would make of the reports and the reader's level of expertise. Further consideration might lead you to arrive at six or forty-nine categories. You might conclude

---

\*    I have adapted this audience material from Professor Wayne Losano's course in technical writing. Professor Losano was my teacher at the University of Florida and has taught these concepts very successfully for many years.

that every person who reads makes up his or her own audience. But for the purpose of explanation, we will assume that there are five categories. It is possible that some of these categories intersect: an operator's manual can be written in both lay and expert terms.

## Lay Audience

The lay person's expertise falls within the realm of common knowledge. This reader possesses no specialized knowledge about the subject. For this reader, avoid jargon (overly technical terms); offer clear and distinguishing definitions. Explain concepts or background material only as they apply to your purpose. Examples, illustrations, or analogies render more abstract ideas into familiar terms and would be appropriate for a lay person. Diagrams, photographs, and relatively straightforward graphics offer the most suitable illustrative material; avoid graphics which require specialized knowledge to interpret them. If you have ever tried to assemble a bicycle from instructions, you know what it is like to be a member of a lay audience. The general rule for this audience is simplify.

## Operator

The operator works with the equipment but is not a theorist. Graphics should be geared to the operator's use. If you drive, your driver's license is an operator's license. The operator needs to know what it takes to use the device or process, just as drivers need to know the rules of the road. Some information about routine maintenance may be appropriate as it is user-oriented. Operating manuals are written for this type of audience.

## Technician

The technician possesses some theoretical knowledge, more at the level of operation than at the level of design. A mechanic knows how to take apart your engine, possesses some theoretical knowledge about the functioning of the internal combustion engine, but probably would not develop a design for a new rotary engine. Graphics are geared to the working parts of the object or process at a level of greater complexity than those for the operator.

## Management

The manager is concerned with the dollars-and-cents aspects of the issue; this person gives the approval to buy, sell, manufacture, or implement. The manager will also translate something esoteric into something practical for a market economy. Because management is interested primarily in practical aspects (hard data is for experts), this reader would probably turn first to the recommendations and budget section of a formal paper. Graphics are geared to this audience's interests.

*Expert*

The expert is well grounded in the theoretical aspects of the subject and on the ramifications of the topic. The expert knows where something might lead, what the social or applied implications might be. The expert possesses a sophisticated level of language, knows the words, has read the book from front to back. Technical language, dense writing, graphics which encode several kinds of information are appropriate. The written text may be subsumed to symbolic or mathematic expressions. The expert will probably be most interested in the results section of formal paper.

New student writers view the "expert" with trepidation. How can a student write with authority to the expert (the professor who has thought about this subject for the last twenty years)? The ability to write to experts requires that students read and understand the salient literature. A thorough and conscientious literature search puts the student on a level field with the expert. Understanding the current literature is the BIG LABOR but every one CAN do it.

Within the category of expert is the peer reviewer. When you submit a formal research paper to a professional journal or when you submit a grant proposal to a granting agency (NSF or NIH, for example), your work may be subject to review by outside readers who are expert in the field. The reviewer assesses the merit of your work and offers a critique. While the reviewer functions in an advisory capacity to the journal or agency, the review is crucial to the success of your writing. Anticipate possible criticism and prevailing views; then incorporate a response to it in your writing. In effect, you will assume a skeptical reader and write defensively. (Further discussion follows in Chapter 6.)

## Diction and Structure

The audience and the subject matter govern word choice (diction). For an audience with equal or greater expertise, choose the language that you are comfortable with. Never write up to your audience if you are not familiar with the language; a reader can recognize when you faking it. Bear in mind that words convey concepts, so the selection of one word over another merits consideration of the concept each word signifies. Take the time to learn if your word/concept usage is appropriate if you are not sure. Then

**Write so that you will not be misunderstood.**

When you make decisions about word choice, be sensitive to the quantity of information carried by the number of words. For example, *bisect* means to cut into two equal segments. *Bisect* expresses in one word what "to cut into two equal segments" takes six words to say. The latter is wordy, meaning that it takes more words to say what might be equally expressed in fewer. "I bisected the circle" is more economical than "I cut the circle into two equal parts."

Not only is the first choice more economical, but, because it carries the same amount of information as the phrase, it might also be said to be more dense. Density refers to a relative ratio of information/words. No absolute standard or normative rule specifies an ideal ratio of information/words; your audience and your purpose govern that decision. Writing that is less dense is writing that conveys information in a greater number of words, relative to a more concise formulation. Subsequent drafts of the same paper tend to become more dense.

> The word *boustrophedonic* means a movement alternating right to left, left to right. It refers to the way that the cattle plow a field. Some computer printers print in this fashion. Which is more compact or dense?

- The ancient Hittites had a style of writing which moved, in alternate lines, from right to left, then left to right, right to left.
- The ancient Hittite style of writing was boustrophedonic.

Both examples convey the same information. Which form is more economical in the number of words? Which form is more dense, compressing more information in fewer words?

Which form is intelligible to all but an educated few? One effect of increased density may be a reduced level of general intelligibility. Sometimes one loses in intelligibility what one gains in economy.

Compare *bisect* and *boustrophedonic*. Both of these words reflect a more economical expression of a descriptive phrase; are both these terms accessible to the same audience?

## Levels of Diction

What general audience level would be appropriate for each of the following words? Use a scale based on relative degrees of expertise with respect to your own.

(1) *Hypsypops rubicundus*      *Garibaldi*      *big orange fish*
  Hypsypops is taxonomic name given in Latin. A more specialized audience would use and understand this term. However, the common name of this fish—what most people call it—is Garibaldi. But some people have never seen a Garibaldi. If you were standing over a tide pool filled with fish and wanted to pick out the Garibaldi, you might point at it and identify the Garibaldi as that "big orange fish."

(2) serrate teeth      saw-toothed      shark's tooth
Serrate means the same thing as saw-toothed, but the latter offers more of a picture. A shark's tooth is an example of a serrate tooth. In terms of relative complexity, serrate teeth is more complicated than shark's tooth because *serrate* abstracts the idea (tooth with ragged edges) from the particular instance of a kind of tooth.

(3) nimbostratus      rain cloud      cloud that makes rain
A *cloud that makes rain* is a phrase that describes the function of the cloud. The description is compressed and assumed when the same idea is expressed as *rain cloud*. For a meteorologist, accustomed to a specialized taxonomy of clouds, the more appropriate term would be *nimbostratus*.

In general, the less experienced your audience, the more descriptive your language will be. This descriptive quality will come from abbreviated explanations in the place of single words, from examples or illustrations as they are appropriate, or from analogies that clarify a point. A less-specialized audience requires you to interpret data for them; this audience will not know what the data mean, so you must explain the meaning. For instance, the implications or significance of experimental results, while clear to a colleague in your field, may require elaboration for someone outside it. The less expertise your audience possesses, the more interpretation and explanation are needed.

The converse also obtains but in a different fashion. Writing usually becomes more dense as you make the transition from a first draft to a final draft. It is common for first drafts to string out the significant meaning. As you revise, go through a sentence at a time in order to isolate the most significant points. Look for a word or words that can most succinctly convey the information. Oftentimes such revision requires you to think out the meaning of your writing and to reconceptualize the phrasing. Consider the following sentences:

- I used only animals with a back bone in my study.
- The study was limited to vertebrates.

In this case, *limit* conveys the same idea as *used only* and *vertebrates* supplants *animals with a back bone*. Manipulating words does not produce this alteration; the whole meaning of the sentence has to be processed and then completely recast.

## Sentence Structure and Audience

Diction or word choice is not the only writing element geared to the audience; sentence structure is also a factor in raising or lowering the density of writing. When you combine, embed, or subordinate sentences, you increase the level of information contained at the sentence level. Compare the following two accounts of the same process. Which account is more dense? Describe the differences between them.

- The unknown substance was dissolved in chloroform. Then it was subjected to Thin Layer Chromatography (TLC). The separated residue was then analyzed under UV light. Then the residue was analyzed under normal light.

- After the unknown substance was dissolved in chloroform and subjected to Thin Layer Chromatography (TLC), the separated residue was examined under two kinds of light: UV light and normal light.

In the first passage, each step in time is represented by a sentence. The events in the sentence are linked together by *then . . . [and] then . . . [and] then.* The process is narratively strung out through four sentences. This pattern echoes that of a laboratory procedure, which would be written giving step-by-step directions.

But in reporting *what was done* (rather than *how to do it*) as in a Materials and Methods section, the writer need not represent the chronology in a step-by-step pattern. It is possible for the writer to respect the chronology of events, but, at the same time, to condense the writing to edit out the step-by-step sequence, as is the case with the second passage. One step in the process was grammatically subordinated in a dependent clause beginning with *After.* Two steps were conjoined by deleting the common (and redundant) elements. The order of elements (UV light and natural light) reflects the order of the steps.

## Diction and Structure

Compare the next two sets of sentences. Identify the audience appropriate for each set.

(1) A fish is an animal with a back bone.
(2) It is "cold-blooded."
(3) Cold-blooded means that its blood is the same temperature as it surroundings.
(4) Fish have gills. Gills are special breathing organs that let the fish take oxygen from the water.

(5) A fish is a cold-blooded vertebrate that lives in water and breathes through gills.

The first set (sentences 1–4) conveys a certain amount of information in several sentences. The second example (sentence 5) conveys the same information but that information is packaged differently; it is made more compact. That information is compacted in two ways:

1. One word does the work of several. *Vertebrate* substitutes for *animal with a back bone*.
2. Knowledge assumed to be known can be deleted. An audience who knows what a vertebrate is would also know what gills are. Sentences (1–4) assumes no such knowledge. Both *vertebrate, cold-blooded,* and *gills* are explained. Sentence (5) does not include those explanations because it is assumed that a person who understood *vertebrate* would also know the meaning of *cold-blooded* and *gills*.
3. Sentences are condensed to omit redundant elements.

**Example: Two Audience Levels**

The following two passages treat the same material, but the first selection is directed to an expert audience and the second to a lay audience. In your own words, analyze the difference between them.

Toxic Elements in Pulp Mill Effluent
Audience: Expert (environmental engineering)
    In recent years, more than thirty organic compounds have been identified as contributors to the toxicity of pulp mill process effluents to fish. Chemical analysis procedures using gas chromatography (gc) have been developed for toxicants in the various waste streams, and the toxicity of effluents to fish has been estimated by summing toxic unit equivalents of the measured toxicant concentrations. For many of the streams studied, good correlation has been obtained between the toxicity calculated from results of gc analysis and values obtained from acute lethal bioassays using rainbow trout.

Audience: General
    A pulp mill uses wood to manufacture paper and its by-products. The useless materials in the papermaking process are carried away by a large flow of water, where it eventually reaches a lake, river, or ocean. The waste flow is referred to as *effluent*.
    In recent years, more than 30 organic compounds which are harmful to fish have been found in papermill effluents. These toxic (poisonous) compounds have been identified with the aid of a gas chromatograph. A gas chromatograph is an instrument that will separate, identify, and record several compounds combined in a liquid. This gives researchers the ability to tell how toxic the stream really is.
    Once the chemicals in the effluent are known, they can be tested on fish

to determine their effect. This sort of test is known as a *bioassay*. It has been found experimentally that bioassays using those compounds found in papermill effluents (using gas chromatography) were deadly to rainbow trout.

## What Is Important?

In a comparison between the two passages, the first question to be asked is, "What does the expert-level passage take for granted?" It assumes the audience possesses a particular vocabulary. *Vocabulary* is not to be taken lightly: Word use in scientific writing relies on strictly demarcated meanings. Further, word use implies familiarity with the *concepts* signified by the words. (Before reading further, try to identify the words the expert-level passage takes for granted.)

The general audience selection defines those terms: It explains the process of papermaking, defines *effluent, toxic, gas chromatography,* and *bioassay.* Because the weight of this passage is given to a definition of terms, the fine points of statistical method implied by the results statement, *good correlation,* are lost.

In both cases, the writer had to determine which concepts were the most important and which concepts could be subordinated. The expert-level version defines a problem (toxicity of pulp mill effluents), describes a method used to measure/model the severity of the problem, and provides the correlation between the model and the problem.

In contrast, the lay passage starts with an explanation of the pulp mill manufacture of paper—the beginning of the paper-making process, and the passage is elaborated through an explanation of the terms. But this passage could equally have started with a description of gas chromatography and how it measures toxicity. It could have started with an explanation of statistical methods used to determine good correlation. It could have started with a history and description of bioassays. A number of starting points existed; the writer had to make a decision about the best starting point in order to accommodate the relevant points. An incorporation of *all* these points would have resulted in an unwieldy piece of writing. The writer made a decision about key ideas to organize and cut accordingly. If every point in the expert passage were to be geared to the lay audience, a number of introductory regressions would have to be made.

In both cases, the packaging of the information was geared to the audience level.

## Extension of the Example

At a surface level, this exercise shows how expert writing translates to lay writing; you've been given two passages with a specified relationship between them and been asked to analyze the relationship. Now, let us take what you have just learned and abstract from it. In the first passage, you were given a very dense piece of writing with much information contained in it; that body of information was repackaged into a writing geared to a different audience. This example shows how writing can be transformed and ideas repackaged.

This exercise is paradigmatic of the broader writing process. In the beginning stages of writing, it is typical and desirable to overwrite, i.e., to collect a body of information which must be reshaped by specific requirements. The body of notes taken from references and your own thoughts constitutes a corpus of information greater and more wordy than the final paper. In making the transition from notes to paper, you are in the position of determining the starting point for your writing, determining the relevant details and information, the concepts which require further elucidation, or the concepts which can be sufficiently rendered in one or two words. What governs this selection is the identified audience and your purpose in writing.

The *treatment* of the subject governs the subject matter; the treatment of the contents is dictated by the writer's purpose and the audience. Shifting levels of writing or revising require more than the mere manipulation of words and structures, each calls for an *identification* of the important points including the patterns and linkages the author wishes to develop, a *selection* of those points based on their relevance, a *reconception* of the meaning of words and ideas, and a *recasting* of the ideas in the appropriate manner.

## Significance of Subordination

*Subordination* is a word that invokes relationship within a hierarchy and denotes the auxiliary, secondary, or less prominent element. Subordination gives information about the structure of priorities. In writing it may refer to the hierarchy of ideas within a paragraph, the hierarchy of ideas within a whole paper, the hierarchy of ideas within a sentence. In its restricted use in grammar, it refers to sentence structure.

Subordination at the structural level governs the weight given to the point in question; changes in subordination at the organizational level produce changes in meaning because the priority of information has been modified.

At the sentence level, the structure of a sentence likewise defines priorities. Main clauses indicate higher priority than subordinate clauses. Subject/verb/object positions carry the weight of the meaning; prepositional phrases, adjectives, adverbs are grammatical places that signify the information is subsidiary to the main grammatical elements. Put the most important information in the most important sentence position.

In the examples given above, the determination of key ideas relevant to the audience governed different points of emphasis. The expert audience, with a greater conceptual background than the lay audience, could accommodate more information, and as such the subordination in that passage was *of a different order* than the subordination in the lay passage.

Keep in mind that altering the density of writing requires more than manipulating diction or sentence structure. Ultimately, there is a *qualitative* level based on the purpose of your writing. Rather than focus on the manipulation of words and structure as an end in themselves, keep foremost in mind that these are means to an end; a person exists at the other end of your communication.

### To Make Your Writing Less Dense
### (for a less specialized audience)

1. Prior to writing, identify the purpose for your writing and its targeted audience. Define the information gap. Determine your priorities: what to include, what to exclude.
2. Use more (relatively) descriptive words—*to cut into two equal parts* or *that orange fish called a Garibaldi*—rather than specialized words: *Hypsypops rubicundus.* (Remember that the difference between descriptive words and specialized words is a relative one; no absolute scale governs this distinction.)
3. Break down difficult concepts and explain them in more simple terms, as they are relevant.
4. Alter your sentence structure, using more simple and more complex sentences than compound/complex ones.
5. Provide the necessary transitions to make the sentences hang together.

### To Make Your Writing More Dense
### (for an audience equal to your level of expertise)

1. Prior to writing, identify the purpose for your writing and its targeted audience. Define the information gap. Determine your priorities: what to include, what to exclude.
2. Per your priorities, compress information in a more economical form. Make one word carry more information than several words. Words like *nimbostratus, Hypsypops, serrate,* or *bisect* contain a great deal of information in a short form.
3. Condense the amount of information contained in sentences by
   a. deleting irrelevant words
   b. deleting or substituting words which carry no informational content: "It is everywhere apparent"

  c. combining sentences using subordination or modification as you
    delete redundant words.
See the Appendix for further explanation of these grammatical processes if
you wish further clarification.

## Exercises

### 1. Audience Identification
Read the following passage on the Ebola virus.[2] Because the writers assumed
that the intended audience would not understand concatemerization, the definition, *coiling*, is provided in parenthesis. Using this definition as a standard to define
a level of difficulty, identify other words of similar complexity that might likewise
merit definition. Discuss.

> The Ebola virus can have different lengths due to concatemerization (coiling).
> However, its average length is 920 nm and its diameter is 80 nm. The Ebola virus is
> basically composed of long filaments essentially of bacilli form and nucleocapsid
> surrounded by a lipoprotein which is most of the time Hydrophilic. Other chemicals
> also make up the biochemical structure of the Ebola virus: the nucleotides that make
> the RNA structure of the virus, phosphates that help the virus go through mem-
> branes, and glycoproteins.

### 2. Changing Audience Levels
a. Read the following passage and identify the targeted audience. Explain your
reasoning.

b. Choose another category of audience and rewrite this description to accommodate the needs of that reader. Explain your reasoning.

#### Fuel Knock and Engine Rating
Knocking is a little understood phenomenon that dissipates considerable energy. It
increases with pressure, and since an engine will develop more power and use less
fuel with an increase in the compression ratio to the maximum point, development
of an efficient high-compression engines is dependent upon the knocking tenden-
cies of available gasoline. The use of *n*-heptane and isooctane (2, 2, 4-trimethylpen-
tane) as standards for rating fuels was introduced in 1927. *n*-Heptane, which is
particularly prone to knocking, is given an arbitrary octane number of 0, and
isooctane, which detonates at only a high compression, by definition has a value of
100. The octane number of a given fuel is defined as the percent of isooctane which
is added to *n*- heptane to produce the same knocking characteristics as the fuel under
examination in a standard single-cylinder engine operated under specific conditions. In the alkane series the octane number decreases as the carbon chain is
lengthened, and increases with branches in the chain. Alkanes have a higher rating
than corresponding alkanes. Cyclic paraffins are less prone to knock than the
corresponding straight-chain compounds, and aliphatic side chains lower the rating. Aromatic hydrocarbons have exceptionally high octane numbers.[3]

### 3. Liver flukes: Explanation
Suppose you are traveling through what used to called the country of Burma,
and you happen upon a farming village of rice farmers. The townspeople are

successful farmers, bringing in a good crop. They have a varied diet, including chicken and pigs raised around the village. Nevertheless, the majority of people suffer from a disease causing constant diarrhea and nausea. A number of children have died from diarrhea. Because much of the work in the flooded rice paddies is done bare-legged and bare foot, you correctly suspect the people are suffering from blood flukes.*

Explain to the people what is happening to them, how they are getting the disease, and what they might do to combat the disease. Remember that these people are intelligent but have no more education in science that a second grade child in this country.

### 4. Analysis of Audience

Chapter 3 opened with NIH grant specifications for an equipment-purchase proposal. Was there any implication about accountability to the public for the use of federal funds in the way the specifications were written?

How would the summary be different if this proposal were written for an expert audience and the award made solely on the merit of the project?

That selection is reproduced below.

> If the project is supported, the Summary will be published by the Foundation to inform the general public about its programs. Accordingly, it should be written so that a scientifically literate lay person could understand the use of Federal funds in support of this project.—NSF, Public Document, Program Guidelines, 1992 (my emphasis).

### 5. Audience Analysis of Journals

a. Go to your library and identify five technical journals in your field. Select three of the five and survey back issues of all three journals for at least one year. What kind of papers or articles does each journal accept or solicit? What do the journals have in common by way of subject matter and writing style? Write a short summary of the content of each journal over a year; the summary should consist of supported generalizations derived from your survey.

b. On the basis of your analysis of the content, infer the audience for the journal and write a description of the audience. (Do not identify the audience simply as "expert" or "general.")

c. The Proceedings of the National Academy of Science, USA, writes in "Information to Contributors" (1993) that "Papers should be written to be understandable to scientists in many disciplines." What would be the common points between papers written for any scientist and papers written for a select scientific audience? You may use the project described in a. (above) as a basis for this answer.

### 6. Shifting Audience Levels

Find an article in a professional or technical journal intended for an expert

---

* The blood fluke (*trematoda*) spends part of its life cycle in snails, the intermediate host where larvae develop before infecting the final host, a human being. The larvae enter the human body by penetrating skin and blood vessels. The flukes mature in the blood vessels of the intestine, after which the eggs pass out of the human host through the feces.

audience. Take the introduction, or a part of it, and rewrite it for an audience with less expertise. Comment on the change in style or levels. Note: This assignment requires more than the manipulation of words and sentence structure. Reshift the conceptual foundation if necessary.

## References

Fieser, Louis F. & Fieser, Mary. *Textbook of Organic Chemistry*. D. C. Heath Co., 1950.
Lannon, J. *Technical Writing*. Little, Brown, and Company, 1979.

## Notes

1. Instructions for PHS 398, OMB No. 0925-0001, 6/30/94, p. 30.
2. "The Ebola Virus," Chan Mainor, Tatiana Segura, Danny Cardenas. Produced for Community College Summer Institute, CAMP, 1995.
3. Fieser, Louis F., and Mary Fieser, *Textbook of Organic Chemistry* (Boston: D.C. Heath Co., 1950), p. 93.

# 4

# Abstracts

An abstract is a self-contained form that exists outside the body of the research paper, and as such, must be able to stand on its own merits, independent of the paper. An abstract *extracts* the most important points of a paper and presents them in a concise way. The abstract uses the same language of the paper. Within a specific word limit defined by the publishing journal or information service, the abstract reports the particulars of the scientific paper. The abstract must include the following information: (1) the issue, (2) how the experiment was done, (3) what the observations were, (4) the meaning of the observations. When appropriate, the abstract might also indicate the significance of the research. Because the abstract is an abbreviated form, one sentence for each of these points may be sufficient.

Notice that each of these four points represents one whole section from the body of the scientific paper. In specifying the issue, the abstract condenses the key point or points of the Introduction. In specifying how the experiment was done, the abstract represents the salient points of the Methods section. In representing the observations following the experiment, the abstract condenses the Results section. And in representing the meaning of the observations, the abstract condenses the important points of the Discussion section. Because the abstract, as a form, is independent of the science paper but uses the same language, it can function as a microcosm of the science paper, and, as such, offers the opportunity for close scrutiny as a prelude to engaging the scientific paper as a whole. For this reason, a separate chapter is devoted to abstracts.

The abstract will be more broadly read than the paper. Abstracts may be published by themselves in journals solely devoted to them. More importantly, electronic information sources will often contain the abstract, which can be subjected to a key word search. The time constraints which might restrict a scientist's interest in a whole paper would make the abstract a desirable form to survey relevant research.

# Indexing and Key Words

Indexing is a service geared to the reader's benefit. Indexes offer a key for a reader to research information. For the benefit of readers, abstracts are published independently of papers; because the abstract can stand alone, it saves scientists time because they can study the abstract for salient points to determine whether or not they should read the paper. To accommodate the burgeoning swell of published information, some journals are devoted solely to publishing abstracts. Such journals must often balance the benefits of timeliness with those of comprehensiveness, sometimes favoring one over the other. At a meeting of the American Chemical Society, for instance, John Lane stated that

> The purpose of *Chemical Abstracts* has been defined as "preparing concise summaries . . . from the *indexing point of view*." Here the goal is to provide a timeless reference tool for chemists. Speed in reporting must necessarily be sacrificed for the sake of comprehensive coverage of the world's literature and for thorough indexing and cross-indexing. . . . At the other extreme, *Current Chemical Papers*, which has replaced *British Abstracts*, cuts appearance time to only a matter of weeks, but at a sacrifice in comprehensiveness. It is primarily an indexed listing of titles.[1]

When you begin the literature search for your topic, your search will be geared to key words in the title of the abstract. In a similar vein, when you write a title, write it from the standpoint of a researcher: Use common key words in your title.

# Two Kinds of Abstracts

Abstracts may be loosely grouped into two categories: the *indicative* and the *informative*. The indicative or descriptive abstract tells what topics were taken up in the report. The informative abstract gives precise information about what the report *says* on those topics. Most abstracts are a combination of both forms.

### Example: Indicative Abstract

(1) Studies of diseases caused by mitochondrial DNA mutations suggest that a variety of degenerative processes may be associated with defects in oxidative phosphorylation (OXPHOS). (2) Application of this hypothesis has provided new insights into such diverse clinical problems as ischemic heart disease, late-onset diabetes, Parkinson's disease, Alzheimer's disease, and aging.—Douglas C.

Wallace, "Mitochondrial Genetics: A Paradigm for Aging and Degenerative Disease?" *Science* 256 (1 May 1992): 628.

This abstract describes a hypothesis (1) and tells you that the application of this hypothesis has given insights into some medical disease (2). But what are these insights? How were they achieved? What was the procedure? The abstract presents the topics but not the specifics of their interaction. This kind of abstract may be the most appropriate to summarize a complex paper in a few number of words.

*Example: Informational Abstract*

(1) The purpose of this experiment is to replicate the studies performed by Spiegel (1979) and Gåssler (1954) and to calculate the critical bandwidth in the auditory system. (2) The tonal complexes of selected bandwidth (40–920 Hz) were constructed digitally using 3–46 pure-tone components spaced 20 Hz apart. (3) For Gåssler's replication, sinusoids were added in 20 Hz decrements from 1100 Hz; for Spiegel's replication, sinusoids were added symmetrically in 20 Hz steps about 1100 Hz. . . . (4) As the bandwidth increased to about 160 Hz, the threshold remained unchanged, thus creating a line with a slope of 0. —Excerpt from Yung-Chia Melissa Wang, "Summation Bandwidth at Threshold for Gåssler and Spiegel Experiment," *Journal of Undergraduate Research in the Biological Sciences* 21 (1991): 741.

This informative abstract provides precise information. Details are specified: (2) (3) (4). This abstract was so detailed, in fact, that only an excerpt was reproduced here. While such detail offers a great boon to the reader, many journals specify word limits which preclude extensive detail.

The decision to write an informative or indicative abstract should be constrained by the audience (implied by the journal to which you will submit it) or by other requirements, as for example, the recommendations by a colleague or your instructor.

# Using Codes in Reading and Writing Abstracts

*Science* magazine gave the following specifications for abstracts:

Abstracts should explain to the general reader why the research was undertaken and *why the results should be viewed as important*. The abstract should convey the paper's main point and outline the results or conclusions. For general articles the abstract should be 50 to 100 words.[2]

**Q. What makes your results important? How do you explain the significance of your research?**

You cannot go into great detail in an abstract, especially one limited to 50–100 words. But you can adopt shorthand codes that a reader will recognize. Scientific criteria for significance are based on (1) recency of research, (2) discovery or innovation framed in the relationship: known to unknown, (3) definition of a problem/solution to the problem.

## (1) Recency

Because science is built upon the idea of progress, one way to show the importance of your work is to emphasize its *recency*. Words which refer to the past have a slightly undesirable connotation. The old way or the traditional point of view is conceived of as being superseded by newer studies. Key words which show pastness imply that the work is dated or outmoded:
- *Previous studies* have shown . . .
- *Earlier work* indicated. . .

Your work, however, supplants the old studies.
- Although (previous studies) have shown some changes, our study shows *new advances*.

Or, perhaps your work is more advanced than even the most recent studies.
- *Contemporary* work shows this, but our work shows *new advances*.

Or, your work, while not exceeding the most contemporary work in the field, fits in with other leading-edges studies.
- Contemporary advances in such and such shows this; our study *builds upon those advances*.

When you read abstracts, look for the past/present arrangement; when writing, adopt this code if it applies to your studies.

### Example: Past to Present

Past experiments using *Drosophila melanogaster* have involved selecting for adult traits, inasmuch as direct selection for larval characteristics has proven less feasible and prone to artifact. *The present study* explores the possible effects of selection forces exerted in the adult state on a pertinent larval trait: *development time.*—Alexander E. Romualdez Olivido, "Larval Development as an Indirect Response Character of Imaginal Selection Regime," *Journal of Undergraduate Research in the Biological Sciences* 21 (1991): 122.

## (2) Known to Unknown

Another way to signal the importance of your work is to indicate that your work makes an innovation; your points will be framed in terms of *Known to Unknown*. The following provides a series of paraphrases for expressing this relationship:

- We know a lot about X, but we do not know everything (why it works, how it got there, how it reacts under ultraviolet light);
- One of the pieces is missing, and we will attempt to fill in the blank;
- We know some of the causes but not all of the causes;
- We understand this in one situation but we do not know how it will behave in another situation.

Note: You need not to reinvent the wheel or the electric light. Every addition to the body of scientific knowledge, no matter how insignificant it may seem to you at the moment, advances the cause of research.

### Example: Known to Unknown

Intra- and extra-cellular matrices are known to maintain cellular rigidity, membrane morphology, and membrane protein topography. Little is known, however, about the components of these matrices that may play a major role in neural connectivity, i.e., synaptic adhesion in the mammalian brain. This project tests for extracellular matrix (ECM) molecules and cytoskeletal components that are positioned in areas of the brain that make them likely to be involved in the formation of adhesive contacts between neurons. —Everard Tesoro Esteban, "Components of Structural Matrices May Help Promote Neural Connectivity in the Brain," *Journal of Undergraduate Research in the Biological Sciences* 21 (1991): 286.

## (3) Problem/Solution

If common agreement exists about social, medical, or scientific problems, you can take the pragmatic approach: There is a problem and you have a solution, or part of a solution, or an attempt to offer a solution. The problem/solution structure obtains in the absence of a consensus. A problem need not be universally recognized as a problem for this strategy to obtain as long as you define the terms of the issue sufficiently to make the case. The abstract may state the problem, or it may assume you already know what the problem is. Sometimes this viewpoint is an indicator about what issues are relevant social issues.

*Example: Problem/Solution*

(1) Insoluble plutonium- and americium-bearing colloidal particles form during simulated weathering of a high-level nuclear waste glass. (2) Nearly 100 percent of the total plutonium and americium in test ground water was concentrated in these submicrometer particles. (3) These results indicate that models of actinide mobility and repository integrity, which assume complete solubility of actinides in ground water, underestimate the potential for adenonucide release into the environment. (4) A colloid-trapping mechanism may be necessary for a waste repository to meet long-term performance specifications.—J. K. Bates, J. P. Bradley, A. Teetsov, C. R. Bradley, M. Buchholtz ten Brink, "Colloid Formations During Waste Form Reaction: Implications for Nuclear Waste Disposal," *Science* 256 (1 May 1992): 649.

If you have located a problem that has not yet been socially or scientifically legitimated—as may be the case in the discovery phase of research—you must provide sufficient detail to demonstrate that a problem exists. Do not be put off by establishment views that say there is no problem. Your job is to explain it.

# Writing Abstracts

Certain factual information must be included in the abstract. The abstract must answer four questions: (1) what was the issue? (2) how was the experiment done? (3) what were the observations? (4) what did the observations mean? Allot one sentence (or more) to each point and provide transitions.

The title of the abstract is very important. The title should be reflect the content of the paper, and it should be succinct and brief. Use key words for indexing and for searches.

*Example: Student Abstract*

"Naturally Occurring Allergens that Cleave DNA"[3]
[Issue] Previous investigations on environmental toxins and die-

tary constitutes have shown that some of the chemicals that we ingest or come in contact with possess the capacity to cleave DNA. [how experiment was done] We have examined the sesquiterpene lactones, known to cause contact dermatitis, for their ability to damage DNA. [what the observations were] In vitro DNA cleavage assays have demonstrated that sesquiterpene lactones possess the capacity to cleave DNA to various degrees. [meaning of observations] Furthermore, from these preliminary findings we suggest that sesquiterpene lactones form free radicals as possible intermediates in the cleavage of DNA.

## Exercises

### 1. Analysis of Sample Abstracts
Analyze the following abstracts using the questions that follow each selection.

### Example 1
(1) American children are worse off than those in the previous generation in several important dimensions of mental, physical, and emotional well-being. (2) During the 1960s cultural changes adversely affected children while material conditions improved substantially. (3) By contrast, material conditions deteriorated in the 1980s, especially among children at the lower end of the income distribution. (4) Public policies to improve the material condition of children require a transfer of resources from households that do not have children to those that do. (5) Government programs such as tax credits and child allowances are more efficient and equitable than employer-mandated programs. —Victor R. Fuchs and Diane M. Reklis, "America's Children: Economic Perspectives and Policy Options," *Science* 255 (3 January 1992): 41.

   a. Is this abstract informative, indicative, or both?
   b. Identify the sentence that states the writer's purpose.
   c. The first sentence contains a comparison: What is the comparison?
   d. What span of time does the study cover?
   e. Look at sentence (4). Is this a conclusion or a recommendation? In other words, is the writer saying this situation EXISTS or is the writer saying this situation SHOULD exist?

### Example 2
(1) Recent experimental results are beginning to limit seriously the theories that can be considered to explain high-temperature superconductivity. (2) The unmistakable observations of a Fermi surface, by several groups and methods, make it the focus of realistic theories of the metallic phases. (3) Data from Angle-resolved photoemissions, positron annihilation, and de Haas–van Alphen experiments are in agreement with band theory predictions, implying that the metallic phases cannot be pictured as doped insulators. (4) The character of the low energy excitations ("quasi-particles"), which

interact strongly with atomic motions, with magnetic fluctuations, and possibly with charge fluctuations, must be sorted out before the superconducting pairing mechanism can be given a microscopic basis. —W. E. Pickett, H. Krakauer, R. E. Cohen, D. J. Singh, "Fermi Surfaces, Fermi Liquids, and High-Temperature Superconductors," *Science*, 255 (3 January 1992): 46

a. What kind of abstract is this?
b. What phenomenon are many researchers trying to explain? (1)
c. What relationship does experimentation have to theory? (1)
d. What aspect of superconductivity will the writers focus on, and why do they choose that aspect? (2)
e. How does data relate to theory? This sentence is like (2) in that it says something about the relationship between experiment and theory. How is sentence (3) different from (1)?
f. What do the writers need to determine to make a more complete analysis?

## Example 3

(1) It is known that some moths and butterflies (*Lepidoptera*) contain venoms either in the larval, pupal, and/or adult stages, which can cause epidemics of contact dermatitis in various parts of the world. (2) The physiological effects of skin contact with these *Lepidoptera* range from skin eruptions to severe pain, hemorrhage, and in some extreme cases, death. (3) The chemistry of the poisons found in toxic butterflies and moths appears to vary from species to species. (4) Very little is known about the toxins, especially their chemical structures, functions, and modes of action. (5) We examined the chemistry and toxicology of the venom from the larval stages of the Costa Rican butterfly, *Heliconius ismenius clarescens*, and the closely related Californian butterfly, *Agraulis vanillae*, whose caterpillars are covered with spines. (6) We describe the isolation of the venoms from these particular species, and show that they are found to be contained within a cavity in the center of the spine. (7) Preliminary chemical analysis of these venoms, by the use of paper chromatography, indicate that the major toxic component of these venoms is histamine. (8) In order to establish the potency of this toxin, hair extracts were prepared and tested using the Guinea Pig Irritancy Test for contact venoms. (9) The two venoms produced very similar reactions to the histamine standard when applied topically to slightly abraded skin. (10) The reaction was the classic inflammatory response consisting of localized redness, flare, and localized oedema.—Rebecca Torres, "The Biology and Chemistry of Lepidoptera Venoms," *Journal of Undergraduate Research in the Biological Sciences* 21 (1991).

a. Mark distinct divisions in this abstract. Where does the introduction end? Where does the description of the experiment begin? Where do the results begin and end?
b. Take the first five sentences and, without substantially changing the content, write them as three sentences. Do not simply combine the sentences.

## 2. Editing an Abstract

The abstract (above) by Rebecca Torres has 234 words. While it is a fine and thorough abstract, the writer wants to submit her article to *Science* magazine. She will need to cut her abstract down from 234 to 50–100 words. You volunteer to act as her editor. Revise the text of the abstract to reduce it by three-quarters to one-half.

## 3. Writing an Abstract

a. Take the "Results" section from a scientific paper and write a 200-word abstract for it.

b. Turn to the Reading Selections at the back of the book and write a 50–150 word abstract to accompany Beijerinck's paper, "*Contagium vivum fluidium* as the Cause of the Mosaic Diseases of Tobacco."

## 4. Analysis of an Abstract

Find an abstract from either a scientific journal or a popular science magazine and explain how the abstract argues for the importance of its subject matter. Then find another abstract on the same topic but from a different journal, and compare and contrast the two.

## 5. Write a Punctuation Rule

Consider the following sentences from abstracts already cited. Why do "plutonium" and "intra" have a hyphen after them? Write a punctuation rule that describes this use of the hyphen.

- Insoluble plutonium- and americium-bearing colloidal particles form during simulated weathering of a high-level nuclear waste glass.
- Intra- and extra-cellular matrices are known to maintain cellular rigidity, membrane morphology, and membrane protein topography.

## 6. Sentence Analysis

You do not always need to know all the words to understand the point that a passage makes. The key to comprehending an otherwise undecipherable passage is to grasp the *relationships* expressed in the sentence, particularly through verbs. For instance, the identity of a particular substance may not be known to you, but you can understand whether or not the substance *increases, decreases, changes phase or form*, or *disappears*.

Paraphrase the following sentences, already cited, emphasizing *relationships* between terms.

TIP: To help you figure out the relationship expressed in the sentence, look for simple grammatical relationships. Because structure expresses thought, the grammatical relationship will tell you something about the relationship expressed in the report. Look for the relationships asserted through the subject-verb-object arrangement of the sentence. You may want to make a separate paraphrase for subordinate clauses or descriptive phrases.

- Nearly 100 percent of the total plutonium and americium in test ground water was concentrated in these submicrometer particles.
- These results indicate that models of actinide mobility and reposi-

tory integrity, which assume complete solubility of actinides in ground water, underestimate the potential for adenonucide release into the environment.

- As the bandwidth increased to about 160 Hz, the threshold remained unchanged, thus creating a line with a slope of 0.
- This project tests for extracellular matrix (ECM) molecules and cytoskeletal components that are positioned in areas of the brain that make them likely to be involved in the formation of adhesive contacts between neurons.
- Data from Angle-resolved photoemissions, positron annihilation, and de Haas–van Alphen experiments are in agreement with band theory predictions, implying that the metallic phases cannot be pictured as doped insulators.
- Preliminary chemical analysis of these venoms, by the use of paper chromatography, indicate that the major toxic component of these venoms is histamine.

## References

Aparcio, Ayde. "Naturally Occurring Allergens that Cleave DNA," *Research Abstracts for Student Presentations* (September 24–27, 1992). National Science Foundation: Directorate for Education and Human Resources.

Bates, J. K., J. P. Bradley, A. Teetsov, C. R. Bradley, and M. B. Brink. "Colloid Formations during Waste Form Reaction: Implications for Nuclear Waste Disposal." *Science* 256 (1992).

Day, Robert A. *How To Write and Publish a Scientific Paper*. ISI Press, 1979.

Esteban, Everard Tesoro. "Components of Structural Matrices May Help Promote Neural Connectivity in the Brain." *Journal of Undergraduate Research in the Biological Sciences* 21 (1991).

Fuchs, Victor R., and Diane M. Reklis. "America's Children: Economic Perspectives and Policy Options." *Science* 255 (1992).

Mills, Gordon, and Walter, John. *Technical Writing*. Holt, Rinehart and Winston, 1978.

Olivido, Alexander E. Romualdez. "Larval Development as an Indirect Response Character of Imaginal Selection Regime." *Journal of Undergraduate Research in the Biological Sciences* 21 (1991).

Torres, Rebecca. "The Biology and Chemistry of Lepidoptera Venoms." *Journal of Undergraduate Research in the Biological Sciences* 21 (1991).

Wallace, Douglas C. "Mitochondrial Genetics: A Paradigm for Aging and Degenerative Disease?" *Science* 256 (1992).

Wang, Yung-Chia Melissa. "Summation Bandwidth at Threshold for Gåssler and Spiegel Experiment." *Journal of Undergraduate Research in the Biological Sciences* 21 (1991).

## Notes

1. John C. Lane, "Digesting for a Multicompany Management Audience,"

Abstracts, 138th National Meeting of the American Chemical Society, September, 1960. Cited in Gordon H. Mills and John A. Walter, *Technical Writing*, 4th ed. (New York: Holt, Rinehart and Winston, 1978), p. 69.

2.  *Science* 259 (1 Jan. 1993): 40. The emphasis is mine.

3.  Ayde Aparcio, "Naturally Occurring Allergens that Cleave DNA," *Research Abstracts for Student Presentations* (September 24–27, 1992). Washington Hilton Hotel, Washington, D.C., National Science Foundation: Directorate for Education and Human Resources, p. 34.

# 5

## Organizing with Headings

Headings are the titles and subtitles inserted in a text to divide up the sections. A title or subtitle is usually a noun phrase or a verbal phrase; it is not a complete sentence. Headings are used in formal papers, in many kinds of studies, and in grant proposals.

Why do you need headings?

Headings demarcate discrete sections of a paper; from a writer's standpoint, they divide the paper into separate writing tasks. Rather than waking up one morning with the prospect of having to write an extensive formal paper or proposal, you can, instead, work on separate sections at different times. When all the parts are assembled, you will have a complete first draft in hand. Start organizing your writing and your thinking in a way that will accommodate headings.

Not only do headings organize the writing task for the writer, they also organize material for the reader. Scientists and researchers must remain current in their fields. Because they devote much time to reading *broadly*, they may not always have time to read *intensively*. Headings direct the reader to the places in which he or she is most interested. For example, a scientist may find a "Results" or "Methods" section more pertinent than an introduction because, having knowledge of the field, that reader would not be attracted to a description of the general context. That reader may have already read the literature or may have written some of the articles surveyed in the literature search. A reader in the management category of audience, lacking expertise or interest in hard science, may turn directly to "Recommendations." Because the manager is less concerned with the *what, how, why* and is more interested in the *application* of that information, that section will be more pertinent.

If you extracted all your headings from the body of a report or a proposal, you would compile the Table of Contents for your report.

Generate headings for a paper or a proposal from the situation to which it responds. For instance, a progress report, which explains changes that have occurred over time, would employ categories like "Past Work," "Present Work," "Future Plans" (or "Work Accomplished" and "Work in Progress"). A proposal, which requests funding, staff, or other support, would employ categories like "Project Description," "Specific Aims," "Significance," among others. The common headings for scientific papers that report results are "Introduction," "Materials and Methods," "Results," and "Discussion." Journals and granting agencies will specify the headings in their directions.

The following discussion will deal with three subdivisions of headings. If you ever write a long or a very detailed report, you may need to come up with more than three levels. It is acceptable to generate more levels as long as your manner of notation is consistent.

## Kinds of Headings

### First-Level Headings

First-level headings divide your report into the major sections. In a scientific paper, they correspond to the Introduction, Materials and Methods, Results, and Discussion sections. First-level headings are written in *all capital* letters. Do not mix caps (capital letters) with quotation marks or with underlining. First-level headings are *centered*. (Some journals observe a different convention for first-level headings, placing them flush-left. Adhere to the conventions of the journal or discipline for which you write.) Place the major heading two spaces above the following text and/or three spaces below the preceding text.

Note: For short papers, you can leave out the "Introduction" heading. It is understood that the title takes the place or fulfills the purpose of "Introduction."

### Second-Level Headings

Second-level headings show subdivisions within a major section. Your introduction, for instance, may include subdivisions like "Previous Research," "Scope of the Study," "Limitations to the Study," "Purpose of the study," and so on. To write second-level headings, use a *combination of lowercase and capital letters*. Unless the conventions of your field specify differently, capitalize the key words in the title. This heading is *flush left*. It may be typographically set apart by bold or underlining.

Start the text on the *next line* and *indent* the first line. The second-level heading is placed two spaces below the preceding text and two spaces above the following text.

## Third-Level Headings

Sometimes subdivisions within a subdivision need to be indicated. Third-level headings show this type of subdivision. For example, under [first-level] "Materials and Methods," within the subdivision "Subjects" [second-level heading], might be found a third level heading—the specific subjects, for instance. Third-level headings resemble second-level headings with the exception that third-level headings are *indented*.

If further subdivisions are needed, a fourth-level heading could be used: Indent the heading five spaces from the left margin. Put a period after the heading, then begin the text two spaces after the period, on the same line.

# Headings and Outlines

If you remember your high school experience in making outlines, you will see that the headings take the form of a traditional outline without the numbers and letters; the outline and headings are structurally alike.

Below is a sample outline using numbers and letters to show the structure and hierarchy of topics.

Effect of Lomefloxacin, Cefazolin, and Cefonicid
on Human Lymphocyte Proliferation
Fernando Thadepalli

I. INTRODUCTION
II. MATERIALS AND METHODS
    A. Antimicrobial agents
    B. Mononuclear cell separation
    C. Mitogens
    D. Mononuclear lymphocyte transformation assays
    E. Measure of [$^3$H] thymidine incorporation
III. RESULTS
    A. Effects of antibiotic addition to unstimulated
        and mitogen-stimulated Mononuclear lymphocytes
    B. Effect of Lomefloxacin on lymphocyte proliferation
    C. Effect of Cefazolin on cell multiplication
    D. Effect of Cefonicid on lymphocyte growth
IV. DISCUSSION
    A. Antibiotic effects on mitogen-stimulated mononuclear lymphocytes
    B. Lomefloxacin and lymphocyte proliferation
    C. Cefazolin on cell multiplication
    D. Cefonicid and lymphocyte growth

Notice that the main divisions in this paper are those of a scientific research paper. If you were to convert this outline to the text of a paper, the phrases keyed to Roman numerals would be first-level headings, the letters second-level headings.

It is important to remember that your selection of appropriate headings draws from the nature and body of your work. While the heads for scientific papers are generally prescribed by the conventions of publication, other writing scenarios may require you to create headings that are appropriate to your work. The following example presents an outline for a collaborative student project on sports injuries to the knee. "Anatomy of the Knee," "Injuries to the Knee," and "Treatment" form the major divisions in the body of the paper.

SPORTS INJURIES TO THE KNEE[1]
Jamale Carlyle, Dustin Cisneros, Premal Patel

I. Introduction
II. Anatomy of the Knee
**III. Specific Injuries**
    **A. Menisci Injuries**
    **B. Collateral Ligament**
    **C. Anterior Cruciate Ligament**
    **D. Posterior Cruciate Ligament**
IV. Treatment
    A. Menisci
    B. Collateral Cruciate Ligament
    C. Anterior Cruciate Ligament
    D. Posterior Cruciate Ligament
V. Bibliography

When the outline is fleshed out with text, the bold segment (Roman numeral III) appears as follows:

SPECIFIC INJURIES
[first-level heading]

As sports and fitness activities become increasingly popular, injuries of the knee occur more frequently. The types of knee injuries vary greatly, ranging from simple contusions and muscle strain to severe injuries at the ligament and joint surfaces. With all the movements that athlete makes during the duration of a game, it is possible that the athlete can injure his or her knee in a mishap. In sports, the most common injuries are related to the knee. Because the knee is such a large joint, various other injuries interact with the knee.

**Menisci Injuries** [second-level heading]

The most frequently occurring injury to the knees is to the menisci. The menisci are common injuries or torn as they become trapped or pinched between the femur condyles and the tibia. The damage sustained by menisci can range from a small tear along the periphery of the cartilage to a more serious injury along the longitudinal tear resulting in a displaced section of the cartilage. Two classic symptoms of meniscal injuries include that of clicking and locking.

**Clicking** [third-level heading]

Clicking is an audible or palpable sensation often caused by torn menisci fragment rubbing against the femoral condyle, while locking is the mechanical blockage of the complete range of motion.

**Locking** [third-level heading]

Locking is usually caused by some kind of internal derangement. The most common cause of locking of the knee results from a fragment of an injured meniscus becoming caught between the femoral condyle and the tibial plateau, thus causing restrictioning of the complete extension. These injuries are the resulting factors of the menisci being injured by all the twisting activities during weight bearing, but can also be caused by direct blows to the knee, or chronic trauma.

**Collateral Ligament** [second-level heading]

The collateral ligaments are most commonly injured by a blow to the lateral aspect of the knee during a sporting event. The collateral ligament injuries are classified as sprains depending on the extent of the tear. A first-degree sprain is a microscopic tear of the ligament fibers that are stretched. A second-degree sprain is also a partial tear, although the injury results in a minute loss of function. A third-degree sprain is a complete ligament tear, with the loss of joint stability. These specific injuries to the collateral ligament are usually common among young adults.

**Anterior Cruciate Ligament** [second-level heading]

The ACL becomes vulnerable to rupture because of its tense stretched position when the knee is rotated or hyperextended. The ACL is frequently torn in contact sports, most commonly in football. These injuries also occur in basketball and baseball when the hyperextended knee is twisted or when the player collides with another player. In football accidents, a forceful blow to the knee often completely tears the ligament. An injury such as this can be a devastating event to the athlete because of the audible pop or crack, followed by swelling of the knee, and the inability to complete the game or continue in the participation of the sport.

**Posterior Cruciate Ligament** [second-level heading]

The PCL is less prone to injury, though a portion of its fibers are utilized through all degrees of flexion of the knee. Rupture of the PCL is much less common than rupture of the ACL and usually results from a direct blow to the tibia. When the PCL is torn, the integrity of the ACL must be carefully evaluated as well. Tears of both the PCL and the ACL indicate a dislocation of the articulations between the condyles of the femur and the tibia.

# Headings in Proposals

In addition to scientific papers, proposals are also organized by headings. The 1992 National Science Foundation (NSF) published grant proposal guidelines for undergraduate science programs seeking desired or necessary equipment to improve courses in science, mathematics, or engineering.[2] The proposal guidelines specified that grants be organized according to a particular outline. The format for the entire proposal was composed of six parts—a combination of written text and completed forms. That format follows:

Proposal Format for 1992 NSF equipment grant
    1. Cover sheet (NSF form 1207)
    2. Project Summary Form (NSF form 1295)
    3. Detailed Budget (Equipment List)
    4. Table of Contents
    **5. Narrative** (Limited to 12 double-spaced pages)
    6. Appendices

Section 5, the Narrative, was to be organized according to the following outline.

**5. Narrative**
    A) The Current situation
    B) The Development Plan
    C) Equipment
        1) The Equipment Request
        2) The Equipment on Hand for the Project
        3) Equipment Maintenance
    D) Faculty Expertise
    E) Dissemination Plan
    F) Citations

## Rules for Headings

1. Check the journal specifications or any special instructions for a preferred style of headings. If no special instructions are given, adopt the form provided by this book.

2. Be consistent in the way you write and punctuate headings. Do not mix punctuation. Do not switch to all lowercased letters in first-level headings if two times out of three, you used all capitals.

3. Be consistent in the divisions and subdivisions indicated by the headings; traditionally, when you write an outline you must have a "B" if you have an "A," or you must have a "2" in the subdivision if you have a "1."

4. You can leave out the "Introduction" heading in a short report. It is understood that the title takes the place or fulfills the purpose of "Introduction."

5. The heading be accurate and complete. The phrase must indicate what the contents of the section will be.

6. Phrase headings in grammatically parallel terms.

## Exercises

### 1. Proposal Outline

The National Science Foundation's Grant Proposal Guide (NSF 94-2, January 1994) provided the following specifications (to be met within fifteen pages) for a part of the proposal. Using a project you have in progress now (or a project you would like to do) as the basis for this question, make an outline of the way you would use those fifteen pages to meet the proposal requirements. What headings and subheadings would you generate?

> The main body of the proposal should be a clear statement of the work to be undertaken and should include: objectives for the period of the proposed work and expected significance; relation to longer term goals of the investigator's project; and relation to the present state of knowledge in the field to work in progress by the investigator under other support and to work in progress elsewhere.

### 2. Results and Discussion

Recall the distinction made in Chapter 1 between *what you see, what you think you see,* and *what you think it means.* In a formal scientific paper, the observations made at the conclusion of an experiment (*what you see/what you think you see*) are placed under the heading "Results" and the significance of the results (*what you think it means*) are placed under the heading "Discussion." In the follow selections taken from "Genetic Variation in the Hawaiian Species *Schiedea* and *Alsinidendron* (*Caryophyllaceae: Alsinoideae*),"[3] match the appropriate paragraphs to the "Results" and "Discussion" headings. Refer to particular words or phrases in each paragraph to support your answer.

> 1. In order to measure the amount of genetic variation, general trends in enzyme variability are noted. The percentage of monomorphic populations for all enzymes resolved is shown in Table 3. Enzymes that were scored for more than one locus (GOT and LAP) usually contained a locus that was much more variable than the other locus. These multiple locus alleles also showed the greatest range in variability. In sixteen of the twenty-one populations analyzed, the fast locus of LAP was the monomorphic, i.e., 76.2% of the populations were monomorphic at this locus. The most variable system of the nine loci surveyed was the fast locus of GOT with only 38.1% of the populations being monomorphic at this locus. The rest of the enzymes have intermediate values between these two extremes. These data show the degree of variability in the enzymes used to estimate the genetic similarity and distance values in *Schiedea*.

> 2. Overall, genetic variability of *Schiedea* is much greater than for other endemic Hawaiian species. Island populations characteristically tend to show smaller amounts of genetic diversity than mainland species due to the high probability of small population size, founder effect, and bottlenecks associated with island colonization. This phenomenon is compounded by additional founder effects as species migrate down the island chain (Witter and Carr, 1987). The great amount of diversity

in *Schiedea* may be attributed to breeding systems that promote outcrossing which is evidence by the lack of variability in autogamous hermaphroditic species. There is no significant difference in the amount of genetic diversity among the outcrossing species because they all contribute to genetic differentiation. Therefore, breeding system does play a role in varying the amount of genetic variability.

### 3. Scrambled Sentence Exercise

The following sentences are taken from a student-written scientific paper.[4] The headings used in this paper include "Introduction," "Materials and Methods," "Results," and "Discussion." Place each of the following statements under the appropriate heading. Generate subheadings as appropriate. Write a title.

1. Compound 1 was eluted from the column using 100% hexane. This component is colorless, does not absorb UV light, is undetectable by sight until applying the elution fraction to TLC using vanillin.

2. Zoopharmacognosy is the scientific study that describes the process by which wild animals select the use of plant-derived properties (i.e., leaves, stems, roots, resins, etc.) with medicinal properties for treatment and prevention of disease.

3. The chemistry suggests insect repellent properties.

4. Sesquiterpene lactone, similar to 8-b-asterolide, has been shown to exhibit ectoparasitic properties. Also, sesquiterpenes similar to b-slinene also have shown to possess insect repellent as well as many other biological properties.

5. Chemical analysis, both technical and analytical, has been utilized as a precursor in understanding the potential therapeutic properties of the resin.

6. The second and third major constituent eluted from the column were collected using a 40/60 hexane/chloroform solvent gradient.

7. Coatis (*Nasua Narica*) of the island of Baro Colorado have been observed utilizing a resin from the tree *Trattinnickia Aspera* as a topical agent.

8. The chemical constituents of the resin include two triterpenes, one terpene, and one sesquiterpene lactone. The chemical constituents suggest ectoparasitic repellency properties.

9. The chemistry of the resin is consistent with the plant family *Burseraceae*. Triterpenes frequently exist in the family *burseraceae*.

10. Bands of coatis approach the grooming tree and scrape the bark with their paws, releasing milky resin from the tree. The milky resin is then groomed into their furs while the brittle-older resin is disregarded.

### 4. Headings with Temporal Divisions

A common way to break up a text is by using an arrangement in time: past, present, future. A progress report, for instance, would make use of these headings, measuring progress by the work completed against the work remaining. Can you think of other writing genres or scenarios that might make use of this temporal division?

## 5. Headings with Divisions based on the Real/Ideal Division

A comparison between the ideal and the real can also be the basis of written projects. How might you use this distinction as the basis of a formal division in a scientific paper?

# References

Day, Robert A. *How To Write and Publish a Scientific Paper*. ISI Press, 1979.

Lannon, John. *Technical Writing*. Little, Brown, and Company, 1979.

Pelayo, Arturo. "The Chemical Study of Resins from *Trattinickia aspera* used by Coatis for Grooming." Unpublished paper produced for Dr. Luis Villarreal's Writing 199 class. University of California, Irvine, Winter 1994.

Straub, Christina. "Genetic Variation in the Hawaiian Species *Schiedea* and *Alsinidendron* (*Caryophyllaceae: Alsinoideae*)." Unpublished paper produced for Dr. Luis Villarreal's Writing 199 class at the University of California, Irvine, Winter 1994.

Thadepalli, Fernando. "Effect of Lomefloxacin, Cefazolin, and Cefonicid on Human Lymphocyte Proliferation." Unpublished paper produced for Dr. Luis Villarreal's Writing 199 class at the University of California, Irvine, Winter 1993.

# Notes

1. Collaborative student paper produced for CAMP (UCI), July 1995.
2. Instrumentation and Laboratory Improvement: Program Announcement and Guidelines, pp. 3–6. Directorate for Education and Human Resources, Division of Undergraduate Education. Closing Date: November 16, 1992. Because guidelines, specifications, and grant awards change, these instructions may be outdated. For further information about applying for actual grants, contact the NSF.
3. A paper produced by Christina Straub for Dr. Luis Villarreal's Writing 199 class at the University of California, Irvine, Winter 1994.
4. Arturo Pelayo, "The Chemical Study of Resins from *Trattinickia aspera* used by Coatis for Grooming." Unpublished paper produced for Dr. Luis Villarreal's Writing 199 class. University of California, Irvine, Winter 1994.

# 6

───────── ✧✦✧ ─────────

# The Scientific Paper

A scientific paper is a major document, of variable length, produced by an individual or a group, which reports original research results, theories, and experimental procedures replicable by peers. The paper concerns itself with new and significant, not trivial issues, of public or professional interest. It includes substantial documentation with references to other work in the field. The paper uses multi-level internal headings appropriate to its subsections, many kinds of graphic representations of data, and formal language. The purpose of the science paper is to publicize data, concepts, and results to an audience composed of one's peers (more below) or that audience typified by the journal.

## Information to Authors

If you write with a specific journal in mind, it is imperative that you locate its "Information to Authors" section. In general, this section provides the content and format specifications for research printed in the journal: It describes the types of articles printed or reviewed and/or the nature of the research; provides specifications on page or word length, number and use of graphics; includes instructions for citing sources; and specifies restrictions or limitations in format or experimentation. Information to Authors can be found by writing to the journal directly or by consulting the journal itself; some journals will print this information in every issue; others will print specifications once a year, say, in the January issue.

*Simultaneous submission* refers to the act of sending out the same paper to different journals at the same time. Most journals will not accept simultaneous submissions; previously published research is also unacceptable in many cases. Consult the Information to Authors for a definition of qualifications to "previously published" to determine if published abstracts or conference papers fall within the boundaries of these restrictions.

# Peer Review

Peer review is a process by which a panel of experts in the field assess the scientific merit of the work. The journal to which you have submitted your paper negotiates the review process. A peer-reviewed paper is reviewed by the editor, an editorial board or an expert *ad hoc* reviewer. Reviewers comment on the manuscript, return the comments to the journal, which considers the recommendations. The reviewers' comments (sometimes anonymously) are returned to the writer with the journal's letter of acceptance, non-acceptance, or provisional acceptance pending revision. For papers with provisional acceptance, revisions must be completed within a specified length of time. If the writer disagrees with the reviewer's criticism, the writer may disregard it, provided that an explanation or justification is offered in the cover letter to the editor. The author of the paper is required to address all criticisms. If the editor feels that the criticism has been treated fairly, either by way of corrected text, new or modified experimental data, or a response in the cover letter, the editor may choose to accept the paper. If the editor believes that the criticism has not been accommodated, the paper may be submitted to a second round of comments.

When writing, consider the perspective of the reviewer: Anticipate potential criticisms and deal with them either experimentally or by presenting an argument. Defend your results and interpretation as if a reviewer were the audience.

What criteria do reviewers employ to evaluate your paper?
The paper may be subject to the following kind of questions.

(1) What is the significance of your work to the field? Does your work extend beyond new observations?
(2) How convincing/clear are the results?
(3) Do the results support the author's conclusions?
(4) How well is the paper written? Does it meet the standards for referenced scholarship?

A general description of the parts of the scientific paper follows. It is intended to clarify terminology, to delineate the parts of the paper, to describe the content, and to recommend an order of writing events. Structural patterns will be recommended and conceptual strategies will be indicated and referenced to other chapters in this book. Specifications in Information to Authors or given by your teacher override these specifications. Sample specifications will be given below.

## Front and Back Matter

By *scientific paper* is meant the substance of the paper itself. Transmittal materials accompany the paper. The title page, the cover letter, the list of figures, and the abstract effectively fall outside the body of the paper proper, and these elements will be termed the *front matter*. *Back matter* is the term given to include bibliography, notes, references, appendices, and glossaries which may accompany the paper. Different disciplines adopt different modes of citing references. To document your sources, consult the standard methods in your field.

Front matter accompanies the formal paper but is outside the full body of the paper itself. It includes the following:

## Title Page

The title page shows:

1. Title of the Paper
2. Serial number if appropriate
3. Writer, writers, or organization publishing paper
4. Person or organization for which the paper was prepared
5. Date of publication or submission
6. Project Number

The title page should adhere to the format preferred by the journal. Some journals eschew the title-subtitle arrangement. Some will limit the number of characters in the title. Do not use a complete sentence for the title. Use key words in the title for ease of retrieval in a computer search and for indexing purposes. If no specifications are given, render the title in full capital letters and center it. Avoid showiness.

## Cover Letter

This letter accompanies the scientific paper and gives the writer the opportunity to make additional remarks not included in the paper. It may explain the significance of the paper, it may note that the paper was invited, it may refer to the subject of the paper and the reason for the paper, or it may simply state the equivalent of "Here is the paper."

The cover letter gives the writer personal contact with the editor or another reader. *Plant Physiology*, for instance, prints three kinds of papers; the shorter (three-page) journal category, "Rapid Communications," presents "a preliminary report or a complete account of significant research contribution." Information to Contributors states that in "the cover letter, the author must explain why the findings merit 'rapid communication.'"[1]

## Abstract

The abstract is a brief section designed to give a thumbnail sketch of the paper. The abstract appears as an independent element, ordinarily not part of the body of the paper. (A summary, in contrast, is part of the body of the paper.) Some journals use abstracting services which publish the abstract separately. See Chapter 4, Abstracts, for more detail.

The abstract must

1. possess a self-contained unity
2. include enough information to satisfy professional requirements
3. be short (one page of abstract for every thirty pages of paper); avoid abstracts over one-half page (c. 150 words)
4. be fluent and readable; abstracts are dense, but some non-information words must be included
5. contain only information discussed in the paper
6. (should) be written after the paper is completed

The *informative abstract* acts as a report in miniature, a capsule version of the report, highlighting main points. It gives the main points (results, conclusions, recommendations), follows the style and the language of the paper, and contains much quantitative information. It is preferred by readers who wish access to the main ideas without reading the paper and by those who have to take action on the paper.

The *indicative abstract* acts as a prose table of contents, indicating the main topics covered in the report. It contains no quantitative information, merely general coverage and development. The nature of the communication and reader preference will determine the type of abstract. For more detail, see Chapter 4.

# Introduction

The introduction contextualizes and defines the writer's research project. The role of the introduction is to bridge the information gap between the writer and the audience. It provides the intellectual context, a statement of the problem, the benefits of the solution, and summarizes previous work on the subject.

Introductions create a context for your study by placing it in a history of related research or in a continuum of contemporary studies which share points of theory, problem definition, procedure, results, or explanation. At a certain point, you will need to include all of the following, but initially focus on the literature search and the definition of the research question. (For further assistance in writing introductions, see Chapter 9, Definition.)

1. statement of the problem: identify the topic
2. background of the problem: other work on the issue
3. the research question you derived (the science question being asked; see Chapter 1)

4. research objective: how you are going to work to address the question; describe how the experiment was done and how it addresses your question.
5. a brief description of observations and conclusions

## Introductions: Examples

Specifications for the introduction will vary from journal to journal, so be sure that your paper frames its parts appropriately. If you unsure about where to begin, start your introduction with a definition (see Chapter 8, Definition).

The following is an example of specifications for the introduction. Notice that the *Journal of Virology* asks for "salient" rather than "exhaustive" references. Other journals, or your instructor, will make different requirements, perhaps preferring a comprehensive literature search to a more restricted one.

### Introduction: Journal Specifications

**Introduction.** The introduction should supply sufficient background information to allow the reader to understand and evaluate the results of the present study without referring to previous publications on the topic. The introduction should also provide a rationale for the present study. Choose references carefully to provide the most salient background rather than an exhaustive review of the topic. Instruction to Authors, *Journal of Virology*, Jan. 1989, p. iii.

### Introduction: Student Example
Effect of Lomefloxacin, Cefazolin, and Cefonicid
on Human Lymphocyte Proliferation
Fernando Thadepalli

(Outline) INTRODUCTION

Antibiotics do not eliminate microorganisms from the body; they can kill microorganisms or retard their growth, thus allowing normal host defenses to clear the organisms and their toxic products. In the absence of adequate concomitant host responses, antibiotics may fail to eradicate an infection. Therefore, the effect of commonly used microbial agents on immune responses may have clinical relevance in many surgical patients. Our investigation examined the effect of a series of antibiotic agents on cell-mediated immunity, using deoxyribonucleic acid (DNA) synthesis by peripheral lymphocytes following stimulation by the mitogens, phytohemagglutinin (PHA) and poke-week mitogen (PWM) as the indexes of cell mediated immunity.

## Writing the Introduction

Start writing the introduction from the perspective of collecting references.

Take notes. In the order of writing events, writing the introduction may come near the end of the drafting process, but you should allow time for several revisions, and you should start the literature search at the beginning of your research. In the introduction, repeat structural themes: what the issue is, how that issue is being addressed. Expect to put in some time doing revisions, shaping and reshaping your writing before the final product emerges with lucidity. Do not underestimate the work involved in writing, particularly that required for the introduction. Do not wait till your experiments are finished to begin writing. Begin writing from the start. Budget your time to allow for multiple revisions. Most manuscripts undergo many drafts.

# Methods

In the Methods section, you report what you did in the course of your experiments. Specific headings emerge from the particular nature of your procedure; in general, headings specify what you did, how you did it, and what happened as a consequence. Typical headings include "Method," "Materials," "Subject," "Procedure."

Students usually begin writing the Methods section first because immediate contact is made with the particular equipment. This section is the most technical and the easiest to write. The purpose of Methods is to provide a comprehensive, consistent, and accurate description of your procedures so another researcher could replicate the experiment(s), calculations, or statistical analysis.

Chapter 1 noted that the experiment was the privileged form of methodology among physical scientists. The Methods section accommodates not only experimental procedures but also statistical or mathematical analysis.

Start by describing the materials, the experimental procedures, and/or the analytic methods you use. Material specifies the substances used; subjects refers to animate subjects. The Methods section includes a series of experiments, calculations, or observations.

## Headings

Use a separate paragraph with a subheading for each experiment or method of analysis. A general rule of thumb may be formulated:

- Use one paragraph with one subheading for each method— whether it be a statistical analysis, description of surgery, a specific process

Arrange these sections in the order that you perform the experiments.

No results are needed at this point in the writing process. Writing Methods is also keyed to the literature search; the literature search is ongoing. Your literature search, in practical terms will extend throughout the time you allot to it.

### Citing Sources for Methods

Some laboratories possess a cook book of common methods and procedures. Methods books contain generic methods or derivative methods. Reviewing original sources of the methods and materials is desirable because, with derivative methods or standardized procedures, the theoretical basis of the experiment is lost. The theoretical basis of the experiment may offer a potential source of experimental error and may merit analysis in the Discussion section.

You cite the source for a procedure in the same way that you cite any other source in your paper. In the first instance (below), the separation procedure and the mass spectrometric procedures have been referenced by a number keyed to the bibliographic details that will follow at the end of the paper. Likewise, the second instance uses numerical notation to reference the way that directional selectivity was assessed.

- Separation procedures for Re and Os have been detailed (20-22). Total analytical blanks, including loading and filament blanks, were approximately 100 pg for Re and 90 pg for Os (23). Mass spectrometric procedures were similar to those described by Creaser and co-workers (24). —M. F. Horan, J. W. Morgan, R. J. Walker, J. N. Grossman, "Rhenium-Osmium Isotope Constraints on the Age of Iron Meteorites," *Science* 255 (28 Feb. 1992): 110.

- I recorded from 111 isolated neurons in extrastriate area MT of three rhesus monkeys and tested each cell for responses to second-order motion. Directional selectivity was assessed with the movement-without-correlation stimulus, a second-order stimulus described by Sterling (17). —Thomas D. Albright, "Form-Cue Invariant Motion Processing in Primate Visual Context," *Science*, 255 (28 Feb. 1992): 1142.

### Methods: Journal Specifications

**Materials and Methods.** The Materials and Methods section should include sufficient technical information to allow the experiments to be repeated. When centrifugation conditions are critical, give enough information to enable another investigator to repeat the procedure: make of centrifuge, model of rotor, temperature, time at maximum speed, and centrifugal force (x g) rather than revolutions per minute). For commonly used materials and methods (e.g., media and protein determinations), a simple reference is sufficient. If several alternative methods are commonly used, it is helpful to identify the method briefly as well as to cite the reference. For example, it is preferable to state "cells were broken by ultrasonic treatment as previously described (9)" rather than to state "cells were broken as previously described (9)." The reader should be allowed to assess the method without constant reference to previous publications. Describe new methods completely and give sources of unusual chemicals, equipment, or microbial strains. When large numbers of microbial strains or mutants are used in a study, include tables identifying the sources and properties of the strains, mutants, bacteriophages, plasmids, etc.

A method, strain, etc., used in only one of several experiments reported in the paper may be described in the Results section or very briefly (one or two sentences) in a table footnote or figure legend.—Instructions to Authors, *Journal of Virology* (Jan. 1989): iii.

### Methods: Student Example
MATERIALS AND METHODS
Electrophoretic data analysis and BIOSYS-1

Numeric designations were assigned to the bands relative to markers whose banding patterns were already known. The electrophoretic data were fed into the computer program BIOSYS-1 (Swofford and Selander, 1989) for analysis. The program generated various genetic variability measures such as allele frequencies, mean heterozygosity, and percentage of polymorphic loci. Similarity and distance coefficients were calculated (according to Nei, 1978) and used for clustering analysis in order to construct a phenogram. Further manipulation of the data was conducted to reveal associations between levels of genetic variability and breeding systems, geographic distribution, and population size, by using existing values and averaging for populations according to breeding system and island distribution. ANOVA's were conducted for statistically comparing the differences in these groups. —Christina Straub, "Genetic Variation in the Hawaiian Species *Schiedea* and *Alsinidendron* (*Caryophyllaceae: Alsinoideae*)," Paper produced for Writing 199, Winter 1994.

# Results

*Results* refers to the direct outcome of an experiment, the figures or values for a set of calculations, or the direct inferences derived from observations. In a causal chain, the results emerge as the effect of a cause or complex of causes. In the Results section, you report "Just the facts."

Graphics, tables, charts, and illustrations are found in this section. Make sure that (1) graphics are cited in the text before the graphic occurs; (2) the graphic is titled and labeled; (3) the graphic is integrated with the text; (4) discussion of the graphic follows the graphic where pertinent. See Chapter 7, Graphics, for an extended discussion.

For every paragraph describing a procedure in the Methods section, you should have one paragraph reporting results.

A general rule of thumb is

- one experiment/one subheading/one graphic

You may lead into your results paragraph by referring back to the procedure. Experiment 1 may produce inconclusive results, in which case a second experiment is devised to supplement Experiment 1. Write up the results of Experiment 1 as they are derived. Do not wait to finish a series of experiments, although you may be in the middle of a series. Record results as they come in. Make your note-taking as organized as possible so you do not end up having to repeat an experiment. Unorganized, scattered detail is not meaningful detail (See Chapter 9, Description.)

### Writing Up Results

Describe results from the standpoint of an impartial observer. Establishing controls can be as difficult, if not more so, than devising the procedure itself, so do not treat them casually. Do not scatter your data throughout the section. An introductory figure that functions as a map may be in order.

For a single experiment or one experiment in a series, provide the following information.

(1) Why did you conduct this experiment?
(2) What did you do?
(3) What did you see?
(4) What does it mean?

Each of these four points will be discussed and examples given below.

(1) Tell the reader *the reason* you are doing this experiment (calculation, etc.). Why did you do it? What purpose did this experiment serve? Provide the rationale.

> In order to determine the molality of . . .

> To test the viability of . . .

As you see, this explanation can be handled in a phrase, a sentence, possibly one or two sentences if the reasoning is complicated. The rationale links the Results section to your Introduction and maintains structural coherence among the separate parts of the paper; in addition, your rationale may merit analysis in the Discussion section and would thus provide a bridge not only between the Introduction and the Results but also between Results and Discussion.

(2) What did you *do*? (one or two sentences)

This part of the Results provides the experimental or analytic response to the rationale.

> We performed gel electrophoresis of six proteins.

> Ten pigmented guinea pigs were anesthetized and . . .

This information will likewise connect the Results to the Methods, to the abstract and Introduction, and to the Discussion.

(3) What did you *see*?

The answer to this question provides the observations which occurred after the experiment ran its course.

> As can be seen in Figure 3, a prominent bend in the gamma chain occurs at the fourth pair.

> Figure 4 shows the weights of reduced copper through six trials.

> The demographic parameters for the spread of disease are shown in Table 1.

Report results. Do not attempt to impose an interpretation on them at this point. A table, graph, or figure may offer the most convenient form to report results; in highly technical papers, a table, calculations, or visuals may supplant narrative results. Sometimes you may need to call attention to important points in the results or to indicate technical limitations to the findings.

(4) What does it *mean*?

> The data *shows* the degree of variability in the enzymes used to estimate the genetic similarity and distances in *Schiedea*.

> This model *reveals* that current estimates of the rate of transmission by macroparasites have underestimated the effect of the full array of factors.

In a most restrictive sense, state specifically what can be concluded or determined. A pattern may be discernable, and it is appropriate to describe the pattern and note the variations in the pattern. You must document any trend you note with the appropriate description of results; do not generalize about data that are not included.

Results may have been inconclusive, in which case, provide the transition to the next procedure and repeat the four-step description for the next set of results. The inconclusiveness of results is itself a conclusion in a logical chain of reasoning, and therefore offers a good place to start discussion in the Discussion section (why were they inconclusive?)

## Results: Journal Specifications

**Results.** In the Results section, include the rationale or design of the experiments as well as the results; reserve extensive interpretation of the results for the Discussion section. Present the results as concisely as possible in *one* of the following: text, table(s), or figure(s). Data in tables (e.g., cpm of radioactivity) should not contain more significant figures than the precision of the measurement allows. Illustrations (particularly photomicrographs and electron micrographs) should be limited to those that are absolutely necessary to show the experimental findings. Number figures and tables in the order in which they are cited in the text, and be sure to cite all figures and tables.—Instruction to Authors, *Journal of Virology* (Jan. 1989): iii.

**Results.** Important results first; data in the form of tables and figures in numerical sequence . . . . Please note that all literature citations belong in the Discussion and not in the Results. Please note also that figures and tables should be referred to parenthetically: e.g., avoid writing "Table 1 gives . . ."—Instructions to Contributors, *The International Journal of Plant Science* 186 (January 1994): 137 ff.

## Results: Student Example

[Reason] In order to determine patterns of variation within and among the 21 different populations of *Schiedea* and *Alsinidendron*, [**What did you do?**] values for genetic similarity and difference were calculated (Nei, 1978) [**What did you see?**] and shown in Table 4. Genetic distance is used to measure the accumulated allele differences per locus between closely related populations (Nei, 1978). The upper triangle in the table shows unbiased genetic identity and the lower triangle shows unbiased genetic distance. Distance is based on genetic identity: $D = -\log_e I$. [**What does it mean?**] The wide range in identity values (0.942 to 0.201) indicates that some species of *Schiedea* are genetically similar, while others are highly divergent.—Christina Straub, "Genetic Variation in the Hawaiian Species *Schiedea* and *Alsinidendron* (*Caryophyllaceae: Alsinoideae*)," Paper produced for Writing 199, Winter 1994.

## Results versus Discussion

Frequently some confusion exists about the relationship between Results and Discussion. *Results* refers to what happened as a consequence of your procedure; *Discussion* refers to the interpretation or the significance of those results. *Results* describes what happened; *Discussion* explains what they mean. *Results* tabulate effects; *Discussion* explains effects.

This difference between results and discussion recapitulates the difference between *what you see* and *what you think it means*.

Results fall within the province of description/observation. They report a clear picture of the outcome. Graphics or visuals are appropriate to the Results section and may supplant written text. Discussion may take the form of a comparison/contrast between expected results and actual results; it may also include a comparison/contrast of your work with other work in the field.

| Results | Discussion |
| --- | --- |
| what you saw | what you think it means |
| observation/ description (see Chapter 8) | interpretation |
| what happened as a consequence of the experiment | the significance of what happened |
| effects | why/how causes produced effects |

## Discussion

The Discussion continues the theme of the introduction by way of identifying and discussing patterns. It interprets results but does not rehash them. The overall flow of the discussion is from particular to general; it should start with a particular statement about the results and then become more general in discussion by referring to the hypothesis or to other studies. In structure it inversely mirrors the introduction, which flows from general to particular. The discussion is generally hard to write; build it like the introduction.

Begin with a general statement about the nature of the results. Then go on to develop your discussion along one of the following lines. The Discussion section of a paper analyzes effects. The following will describe three ways to organize the Results section, none of which is meant to impugn alternative ways of structuring results. Because these are organizational structures that organize the same results, some overlap will be apparent. The three ways include: (1) comparison/contrast with Results; (2) discussion of results organized by the topics of Structure, Function, and Evolution; (3) assessment of the quality of the evidence. These are not the only ways to structure a Discussion.

Make sure that you revise the Introduction in light of the organization of the discussion to preserve structural consistency by engaging the same topics, in the same order, in both parts.

## (1) Comparison/Contrast

The effect is the terminal point in a process of causal relations. To discuss results, compare and contrast the end (result) with respect to the beginning (hypothesis) in order to account for the change by an explication of the event or phenomena that intervened to make the end result from the beginning.

1. Start the Discussion with a reiteration or summary of the results. Make a general statement in the Discussion about the nature of the whole set of observations. Connect this statement to points raised in the Introduction. Elaborate what is clear in these points. Integrate the literature in order to compare results to what others have done.

2. The last point of each section in the results (sentence 4, below, results") is the clearest point of each observation; it states conclusive findings. Starting with this sentence, compare and contrast it with your hypothesis (c/c anticipated results with actual results); with other work in the field (c/c your results with others' results). Assess the results; account for your findings. Limitations to the methods?

3. Repeat this procedure for the results of subsequent procedures.

4. Pull your discussion points together in a final paragraph, possibly leading to recommendations or some moderate speculation based on the rest of your paper.

## Maintain Structural Coherence

The parts of your scientific paper must possess an integrity in and of themselves, offering full and comprehensive descriptions or explanations as they pertain. At the same time the different parts must maintain unity as a body. At some point in the writing—revising the complete first draft—the parts must be interconnected to give the paper structural coherence. A repetition of themes will be the linking content to these parts. To give structural coherence, maintain the same order of discussion in the Methods, Results, and Discussion sections, as follows.

Materials and Methods
Procedure 1. Description of procedure
Procedure 2. Description of procedure

Results
Procedure 1: Results discussed as follows:
 1. reason for the procedure

    2. what you did
    3. what you saw
    4. what it means
Procedure 2: Results discussed as follows:
    1. reason for the procedure
    2. what you did
    3. what you saw
    4. what it means

Discussion

Procedure 1: begin with sentence 4 from results
    c/c anticipated results (hypothesis) with actual results
    c/c with other results (literature search)
Procedure 2: begin with sentence 4 results
    c/c anticipated results (hypothesis) with actual results
    c/c with other results (literature search)

## (2) *Structure, Function, Evolution*

This pattern could produce a five+ paragraph discussion: one paragraph is allotted to a reiteration of the results; one paragraph is allotted to each of the three topics; and one paragraph is allotted to an overview of the previous four paragraphs. Five is not a fixed number; generate the number of paragraphs from your work.

1. Start the discussion with a reiteration or summary of the results. Make a general statement in the discussion about the nature of the whole set of observations. Connect this statement to points raised in the Introduction. Elaborate what is clear in about these points. Integrate the literature in order to compare results to what others have done. Generalize your results, going back to points raised in the Introduction.

2. Discuss your results with respect to structure, function, and evolution (one paragraph each, minimally). *Structure* refers to formal elements: *function* refers to operation; *evolution* or *growth* invokes a larger temporal context.

3. Pull your discussions points together in a final paragraph, possibly leading to recommendations or some moderate speculation based on the rest of your paper.

For instance, if your results describe sickle-cell frequency in a select population, those results might be treated per the tripartite division as follows: The *structure* paragraph could describe the *structure* of the recessive allele responsible for sickle-cell anemia. It might note that abnormal hemoglobin, caused by the substitution of a single amino acid in the protein of the hemoglobin, causes these molecules to stick together, and that a single copy of the recessive allele enhances individual resistance to malaria. The *function* paragraph could describe the *manifestation* of the genotype, possibly comparing/contrasting the behavior of the disease in heterozygous and homozygous individuals. This section might be related back to structure by indicating that, in heterozygous individuals, the two alleles are codominant at the molecular level, producing both normal and abnormal hemoglobin. The *evolution* or growth paragraph might discuss *statistical distribution* of the disease in a select population, analyze gene frequency within a stipulated population, possibly predict gene frequency or spread of the disease. Citation of previous studies threaded throughout the discussion is appropriate.

## (3) Significance of Results

After a summary of the results, the following questions can be addressed in one paragraph per question.

1. What is clear from the results?
    Usually a type of pattern can be defined.
2. What are the limitations of the method?
    Evaluate the methods.
3. What do the results argue for or against?
    Consider negative results.
    Present arguments for both sides.
4. What are alternative explanations? If you favor one, explain why the others are not favored.
    This is the place to anticipate reviewer's criticism.
5. What do the results of others/previous research have to say about these issues?
6. What is the overall or general significance or recommendation of these results?

### Discussion: Journal Specifications
**Discussion.** The Discussion should provide an interpretation of the results in relation to previously published work and to the experimental system at hand and should not contain extensive repetition of the Results section or reiteration of the introduction. In short papers, the Results and Discussion sections may be combined. Instruction to Authors, *Journal of Virology* (Jan. 1989): iv.

### Discussion: Student Example
Overall, genetic variability of *Schiedea* is much greater than for other

endemic Hawaiian species. Island populations characteristically tend to show smaller amounts of genetic diversity than mainland species due to the high probability of small population size, founder effect, and bottlenecks associated with island colonization. This phenomenon is compounded by additional founder effects as species migrate down the island chain (Witter and Carr, 1987). The great amount of diversity in *Schiedea* may be attributed to breeding systems that promote outcrossing which is evidence by the lack of variability in autogamous hermaphroditic species. There is no significant difference in the amount of genetic diversity among the outcrossing species because they all contribute to genetic differentiation. Therefore, breeding system does play a role in varying the amount of genetic variability.

# The Process of Writing: Literature Search

## Defining the Boundaries

A literature search must be comprehensive in that it must encompass the most relevant research pertinent to your topic. Perhaps the most difficult part of this search is your defining the boundaries of relevant studies pertaining to your work. This chore often seems overwhelming to beginning writers. Identifying your audience and understanding the purpose of your writing will help you define the parameters of your search.

If your audience is a peer group, define your peers. You need not consider your social peers, i.e., the other students in the class, as your intellectual peers. If your research has been highly specialized and intensive, other researchers in the field, rather than your classmates, may serve as your audience. Assess the group with which your work has the most in common.

If you write with a specific journal in mind, your familiarity with the papers it regularly prints will help you limit the borders of the subject. The targeted journal will prescribe the context; within that context, assess the information gap between your knowledge and the audience's knowledge. For example, if your research project compared the gamma chain in equine, bovine, primate, and human fibrinogen in order to provide insight into the evolution of fibrinogen, the question to pose to yourself is this: How much evolutionary theory must be provided in the Introduction? Must you go into depth on Darwin and evolutionary theory? Is it possible to truncate the history of the issue to accommodate only contemporary research?

If your audience is familiar with the fundamental theory of evolutionary biology, you need treat this subject only with regard to its more contemporary variations, especially in terms of the way that the salient features of background information have been consistently rendered in the targeted journal. If the journal is geared to a lay audience with passing knowledge of Darwin, fill in the information gap by providing sufficient background.

If you write with no journal in mind, assume the audience to be that of your (self-defined) peers.

## Integrating Sources

Comparison/contrast and classification are the modes that facilitate the integration of other sources with your work. An extensive comparison/contrast of every work with yours is probably not in order. Touch on what is relevant to your study. Compare and contrast others' work with your own on the basis of hypothesis, experimentation, and results. See Chapter 11, Comparison/Contrast.

## Compiling a Body of References

### 1. Use key words in the search

Subject searches are organized by key words in card catalogues and computer searches. Use key words to initiate your search. If you are unable to locate the subject you need, check with the librarian or consult a directory of key words.

Do not limit yourself to a search based on one key word. Extend key words to include synonyms or related words. Generate related words from the theoretical foundation of your paper, from processes, methods, or subjects or models. Build a preliminary database. If an author appears frequently in the subject search, check to see if this author has a larger body of work which may be relevant to your study but which may not be elicited by the key word searches.

Professional journals use indexes to list names and titles of research. These indexes are usually published once a year and index the work in the previous year. If an index is not available for the year you seek (the current year), look through the table of contents in each issue of the journal for the year.

Do not assume too early in the process that your search is complete. The literature search is ongoing. Keep at it. Eventually a common set of references will become clear.

### 2. Assemble a body of texts

Ideally, the separate searches should produce common works. Look for a body of common references. If several papers cite a common source, look up that common source; it could be a seminal paper on the topic and must be included in your documentation.

Make a list of the references. Because much time will be devoted to reading, make a reasonable judgment based on the title, abstract, or introduction about the works most pertinent to your research. You may want to read the most pertinent works first and then engage the others.

Do not skip articles which seem to have limited relevance; read them in order to determine their relevance.

Unfortunately, not all the references you find will apply to your research interest. Nonetheless, this labor-intensive task cannot be over-looked in the event that you pass over an article crucial to your studies. Through the literature search, you begin to create your credibility as an author and build the foundation for your reader's trust.

### 3. Read all the works in your corpus

While all works you have collected in your research may not be important to your writing, you must read through all the works to have an exhaustive search. Make note of seminal papers that are frequently cited in the literature. Check the references at the end of the papers against your list of references and modify your list, renewing your search through subsequent returns to the library or the computer database. (The literature search is ongoing.)

### 4. Take notes

Keep note cards with the reference information, title, and a brief summary of the article. If you have devised your own system for note-taking, use that. List all the referencing information in the event that, as you begin to write other sections of the paper, an unexpected omission in your note-taking requires your rereading some pieces already considered.

### 5. Classify the articles/authors you read

As you take notes in the reading, organize a system of classification. Grouping your references into categories will make for ease of writing the paper later. Define categories relevant to your research, but let those categories emerge from your research. As you come to write, you will integrate your literature search by grouping the authors with respect to their positions vis-à-vis your topics. Although the articles will initially strike you as dissociated names at the beginning of your search, they will later come to stand for different positions as you start to write: support for your view, an alternative view, an exception, an explanation.

For example, in researching the role of the male and female function in hermaphroditic plants, with the intention to examine the role of flowering in such plants, you read ten articles on the subject.

Among the ten you have read, four are relevant to Sub-topic A (the excess of flowers relative to fruit production); the other six are relevant to Sub-topic B (descriptions of flowering patterns). Because you are interested in Sub-topic A, not B, you focus your attention on A.

The four relevant to Sub-topic A can be further divided: Three papers maintain the flowers attract pollinators; one paper maintains that flowers are focused on the male component of fitness. As you integrate these categories in the introduction, they will appear as follows:

(1) Some biologists have suggested that the excess of flowers serve to attract pollinators (Wilson and Rathke, 1974; Willson and Price, 1977; Stephenson, 1979). (2) In contrast, other studies have shown that this floral characteristic has been specifically selected for the male component of fitness (Queller, 1983).

The contrasting point of view (2) is reflected by only one person. But this alternative perspective is not discounted because only one person subscribes to it; rather, this individual is treated as a class of explanations offering an alternative to the more widely held view and, as such, comprises a set with one member.

## Drafts

Expect to write several drafts. It is best to write the first draft in lengthy form. Overwrite. Assume it will be cumbersome and too long. Do not get stuck or blocked in one section. Put down enough to retain your train of thought even though you may be tentative about what you are writing. It may become clear that figures need to be redone.

In the second or third draft, revise for structural consistency. Make sure that the topics in the Introduction are commented upon elsewhere and that all the data is discussed in the Discussion. Compare and contrast Results and Discussion sections to ensure that all the data described in the Results is accounted for in the Discussion. Every point in the Discussion must be raised earlier in the paper. Do not discuss any results for which you have provided no data. Arrange the order of the points.

Beginning in the second and moving into the third draft, begin to revise at the language level to heighten clarity and brevity and to reduce redundancy. At the same time, revise for cohesion among the parts. You may require more than two or three drafts.

By the final draft, the writing should be concise and clear; the physical presentation of your paper should be neat and orderly; the text of the paper should be complete, with visuals and references assembled and properly located in the paper. Drafts always get better.

The following sections are taken from a student-written manuscript. They illustrate the difference between a first draft and a final draft. Notice that the first draft and the final draft differ not only quantitatively but qualitatively. In the first draft, the general parameters of the research are described; in the course of revision, actual experimentation, and research, the general ideas become fleshed out with detail to such a degree that the final draft seem scarcely recognizable from the first.

## Exercises

### 1. Comparison of Drafts
Compare and contrast the first draft with the final draft in the following student paper. How has the final draft changed structurally? In content? In style?

## Effect of Lomefloxacin, Cefazolin, and Cefonicid
## on Human Lymphocyte Proliferation
### Fernando Thadepalli

### (Outline) INTRODUCTION
Antibiotics do not eliminate microorganisms from the body; they can kill microorganisms or retard their growth, thus allowing normal host defenses to clear the organisms and their toxic products. In the absences of adequate concomitant host responses, antibiotics may fail to eradicate an infection. Therefore, the effect of commonly used microbial agents on immune responses may have clinical relevance in many surgical patients. Our investigation examined the effect of a series of antibiotic agents on cell-mediated immunity, using deoxyribonucleic acid (DNA) synthesis by peripheral lymphocytes following stimulation by the mitogens, phytohemagglutinin (PHA) and poke-week mitogen (PWM) as the indexes of cell mediated immunity.

### (Final Draft) INTRODUCTION
The immunoglobulating effects of several classes of antimicrobial agents have been studied by a number of investigators (1, 2, 4, 6, 7, 12, 14, 16). Cephalosporins are commonly used, wide-spectrum antimicrobial antibiotics that inhibit bacterial cell wall synthesis (6). Quinolones are the new age of antimicrobial therapy which target cellular DNA topoisomerase II in S and $G_2/M$ phrases of the cell cycle (13, 18, 22, 23). The expanded use of cephalosporins and quinolones during the past decade has generated a number of chemically modified derivatives that are generally classified by "generations." The molecules of each generation have similar microbial activity but different clinical and pharmacological features (half-life, serum peak concentration, side-effects, etc.) (2, 11, 25). Beside microbial activity, some cephalosporins and most quinolones show immunodepressive effects on the cellular and/or humoral arms of the immune response (2). In particular, a dose dependent inhibition of lymphocyte proliferation in response to mitogens such as PHA, PWM, and con A has been repeatedly observed (2, 11). The degree of such inhibition is variable, depending on the different cephalosporin or quinolone, and mechanism of the depressive effect is largely unexplored. In order to elucidate the cellular and humoral components of the cephalosporin or quinolone-mediated lymphocyte inhibition, we chose to study two cephalosporins, namely, cefazolin and cefonicid, and the quinolone, lomefloxacin. We evaluated the proliferative responses in the presence or absence of the antibiotic, following stimulation with phytohemagglutinin (PHA) and poke-weed mitogen (PWM). This panel of stimuli was selected be-

cause response to PHA and PWM would show independent stimulation by T and B cells, respectively. Our study showed that neither lomefloxacin nor cefazolin significantly depressed lymphocyte proliferation, whereas cefonicid exhibited a dose-dependent depression of triated thymidine uptake in unstimulated and phytohemagglutinin-stimulated MNLs (T cells).

\*\*\*

### (Outline) DISCUSSION

Most antibacterial drugs do not affect the mitogenic response of human T and B lymphocytes, whereas a few are inhibitory. Therefore, the finding of a pronounced alteration in the incorporation of triated thymidine in the presence of cefonicid was remarkable. A similar view is given by the lack of an effective inhibition of proliferation by cefazolin and the quinolone, lomefloxacin, as these might represent formidable drugs which may be used in the immunocompromised patient.

### (Final Draft) DISCUSSION

The results of these studies demonstrate that cephalosporin, cefazolin, and the new quinolone lomefloxacin did not cause a general depression of response in unstimulated and mitogen-stimulated MNLs, whereas the cephalosporin cefonicid did cause a dose-related depression of transformation response in unstimulated and mitogen-stimulated MNLs.

Most antibacterial drugs do not affect the mitogenic response of human T and B lymphocytes, whereas a few are inhibitory (2, 6). Lomefloxacin is a new quinolone that neither affects mitogen stimulated or nonstimulated T and B lymphocytes where cultured for periods of 72 hours. Assays of 72 hours or more revealed a depression at concentrations of 25 and 50 ug/ml but no response was viewed at levels below these, that is, at concentrations that are therapeutic serum levels. This finding suggests that lomefloxacin could be administered to the immunocompromised patient and might not interfere with the generation of an effective immune response by the host.

Previously, Chaperon and Sanders (6) reported that several cephalosporins, including cephalothin, cephradine, cephalexin, cefamandole, and cefoxitin, produced a dose-related reduction in lymphocyte response to phytohemagglutinin and poke-weed mitogen. With the exception of cefamadole and cefalothin, these effects occurred at concentrations that exceed therapeutic serum levels. In the present study, the concentration of cefazolin, which caused a depression of MNL responsiveness, are not achievable therapeutically as well. Cefazolin, like lomefloxacin, is an antimicrobial element that may have clinical significance when we choose to use an antibiotic that does not adversely affect host immune response to treat gram negative infections in the immunocompromised patient.

Due to the different kinetics of the proliferative response induced by two stimulators used in this study, the relative sensitivity to Cefonicid induced inhibition varies with time of culture. Thus at three days the greatest suppressive effect is shown by the mitogen PHA whose responses are severely reduced not only by the highest (200 mg/l), but also by the lower concentrations tested (25, 50, 100 mg/l). Our study demonstrates that *in vitro* exposure of human PBMC to certain concentrations of Cefonicid can result in the depression of lymphocyte proliferation. The mechanism of observed depression is unclear. Cell viability studies in which the tryphan blue exclusion method was used revealed that cefonicid cells did not affect viability at the concentrations used in the experiment. An intracellular effect by the antibiotic is possible, but there is no data available on the penetration of cefonicid into cells. The clinical consequences of the depression of MNL transformations remains to be determined. In situations in which short-term antibiotic therapy is being used for acute infections, adverse effects on host defenses are unlikely. However, in situations in which long-term antibiotic therapy is required, the immunosuppressant effects observed may be clinically important.

## 2. Complete Student-Authored Paper
Read the following paper carefully, and then answer the following questions.
a. What is the purpose of this study?
b. How did the experiment address the purpose? Assess the controls.
c. Outline the Results section, then compare and contrast it with the Discussion and Methods sections.
d. Highlight the numerical references in the author's text. How did Cesar Zamora use references?

## Phototoxicity Comparison of Phthalocyanines
### Cesar Zamora

ABSTRACT

The phototoxicity of Al phthalocyanine tetrasulfonate obtained from Ciba Geigy and Porphyrin Products were compared using a cloning assay. Dark toxicity studies carried out at varying peaks were obtained by adjusting the photosensitizer's concentration utilizing data acquired from spectral graphics. Chinese Hamster Ovary cells (CHO) were plated out, and drug from both sources was added at different concentrations. An argon-dye laser, emitting light at 675 nm and a power density of 50 mW/cm$^2$, was used for the irradiations. The surviving clones were stained and counted. Data shows no clone survival except for the Ciba Geigy photosensitizer at a drug concentration of 1.0 ug/ml. The results indicate that there is higher number of clone survival for irradiated cells at all concentrations drugged with the photosensitizer manufactured by Porphyrin Products than with the photosensitizer manufactured by Ciba Geigy. The results also indicate that there are no cytotoxicity effects for either photosensitizer for concentrations up to 100 ug/ml.

INTRODUCTION

Photodynamic Therapy (PDT) is a promising modern experimental technique used for the treatment of various malignancies (Dahlman, et al., 1983; Waldow, et al., 1987; Dougherty, 1984, 1987). The process of PDT involves the utilization of a photosensitizing dye which selectively localizes in cancerous tissue (Dougherty, 1974); the tumor is destroyed when laser light with sufficient energy is tuned to the appropriate wave-length of the photosensitizer (Gomet, et al., 1989). When photosensitizer molecules become excited by the energy of the laser light, high reactive singlet oxygen is generated (Kimel, et al., 1989). Singlet oxygen plays a major role in the tumor destruction of PDT by interacting with cellular components and tumor vasculature (Gal D., 1992).

PDT has been applied for the treatment of specific tumors (Dahlman, et al, 1983; Waldow, et al., 1987; Li, et al, 1990); Photofrin II which absorbs at 630 nm is the primary photosensitizer utilized in clinical procedures (Kimel, et al., 1989). Longer wavelengths allow deeper light penetration into a tumor; therefore, photosensitizers absorbing at longer wavelengths than PfII would be more advantageous in clinical PDT (Roberts, et al., 1989). Patients must remain out of the sunlight for approximately one month to avoid harmful reactions (Roberts, et al., 1988). Reports of the consequences of PDT treatment have not indicated any permanent side effects of various drug treatments (McCaughan, 1990), but there are possible harmful side effects of various drug treatments. (Roberts, et al., 1989; McCaughan, 1990; Salet, Christian, et al., 1989). One of the most common side effects is skin photosensitivity (Dougherty, 1986), which may be alleviated by a process known as "photobleaching" (Mang, et al., 1987; Potter, et al., 1987; Boyle and Potter, 1987).

The purpose of this study was to establish a comparative phototoxicity of the same photosensitizer, Al phthalocyanine tetrasulfonate, which we obtained from different commercial sources, Ciba Geigy and Porphyrin Products. Phthalocyanines are presumed to be one of the most promising photosensitizers in PDT (Savistsky, et al., 1992). Al phthalocyanine tetrasulfonate absorbs at a wavelength of 675 nm, which will be the wavelength of the laser light during irradiation of cells.

A dark toxicity study, to determine any toxic effects on cells without laser irradiation (Roberts, et al., 1989), will also be carried out. Dark toxicity studies will be performed in conjunction with laser irradiation using the same phototoxic drug concentrations, and a separate experiment for each photosensitizer will also be carried out at much higher drug concentrations, up to 100 ug/ml.

The overall scope of the study was to determine which of the two, the photosensitizer manufactured by Ciba Geigy or Porphyrin Products, had the highest phototoxicity level.

MATERIALS AND METHODS

Absorption Analysis

Chinese Hamster Ovary cells (CHO) were used. These cells are convenient to work with due to their high cloning efficiency and action of dihydroxyanthraquinone on the cell cycle progression and survival (Traganos, et al., 1980). These cells divide approximately every 12 hours. Before commencement, the absorption peaks of Al phthalocyanine tetrasulfonate manufactured by Ciba Geigy and Porphyrin Products were analyzed with a spectrophotometer. Analysis of the photosensitizers indicated that the Porphyrin Products had higher absorption peaks than the Ciba Geigy photosensitizer. At the wavelength of 675 nm, the drug manufactured by

Porphyrin Products had a higher absorption peak than the drug manufactured by Ciba Geigy. Therefore, the Porphyrin Products' drug was adjusted by adding Phosphate Buffered Saline (PBS) solution to obtain equivalent absorption peaks. The use of the presumed drug concentration, without absorption analysis and concentration adjustment, would likely result with a higher phototoxicity for the drug with the highest absorption peak.

Cell Culture System

Subculturing was performed prior to experimentation. The purpose of subculturing is to prevent the cells from exceeding the capacity of the medium, which may cause the cells to cease or greatly reduce their growth. The cells divided until they were approximately 80% confluent, after which they were subcultured. Depending on cell type, a specific enzyme added to the cells to will detach them from the bottom of a flask. In this case, 1 ml of trypsin was added, and the cells were placed in an incubator for approximately 5 minutes to allow complete detachment. The cells were then removed from the flask and added into a centrifuge tube, which was placed into a centrifuge machine and spun at 1,000 rpm for 5 minutes. The supernatant was removed and the pellet was resuspended with media. The cells were then distributed into three different flasks, at specific concentrations, and placed in an incubator with 5% carbon dioxide.

Dark Toxicity

A dark toxicity study was carried out simultaneously with the phototoxicity study at the same drug concentrations. Therefore, enough dishes were plated out for both studies. Dark toxicity studies determine whether or not photosensitizers have cytotoxic effects on cells without laser irradiation. A dark toxicity study with extremely high drug concentrations was also performed for both photosensitizer. This study involved drug concentrations of 0 ug/ml, 10 ug/ml, 20 ug/ml, 40 ug/ml, 80 ug/ml, and 100 ug/ml.

Phototoxicity Assay

Cell phototoxicity was determined by a process known as a cloning assay (Roberts, et al., 1988). After the cells acquired an 80–90% of confluency, trypsin was added to the flask to detach the cells. The cells were then added into a 15 ml centrifuge tube and spun at 1,000 rpm for 5 minutes. The supernatant was removed, and the pellet (the CHO cells) was resuspended with 4 ml of media, and was thoroughly mixed. The cells were then counted using a hemocytometer, an instrument used to count the number of cells for each milliliter of solution. After the cells were counted, 250 cells for every petri dish were plated out (six petri dishes were plated out for each drug concentration and for each photosensitizer). The cells were then allowed to settle and attach to the petri dishes for 2–4 hours in an incubator. During this time period, preparation of different drug concentrations followed.

The preparation of different drug concentrations was carried out under light-restricted conditions. The different drug concentrations included 0 ug/ml, .05 ug/ml, .1 ug/ml, .25 ug/ml, .5 ug/ml, and 1.0 ug/ml. Depending on the concentration desired, a specific amount of drug was added, with a micro pipette, to every ml of Minimum Essential Medium (MEM). Enough drug was prepared to perform both the phototoxic and the dark toxicity experiments simultaneously. Half of the dishes were used for Ciba Geigy and Porphyrin Products phototoxicity study, and the other half were used for the dark toxicity study.

After the cells adhered to the bottom of the dishes, the old media was removed and 4 ml of each drug at a specific concentration was added to six separate petri dishes (This step and the following ones were also done under light restricted conditions.). Three dishes were used for the photo-toxicity study, and the remaining three were left as dark toxicity controls. The cells were then placed into an incubator with 5% carbon dioxide for a period of 18 hours. Afterwards, the drug was removed and 4 ml of fresh MEM was added in its place. The dishes were then placed into the incubator for another 2 hours, and were subsequently ready for irradiation. The same procedure, as mentioned above, was carried out for the dark toxicity study at drug concentrations ranging from 0 ug/ml to 100 ug/ml.

Laser System

An argon-dye pumped laser was used. The laser was set to emit laser light at a wavelength of 675 nm. The power was set at 1414 mW, the energy density was 10 J/cm$^2$, the spot size of the light was set at 6 cm, and the power density was 50 mW/cm$^2$.

Irradiation

Three of the six dishes for each drug were irradiated for 3 minutes and 20 seconds, to obtain an energy density of 10 J/cm$^2$. The remaining half of the dishes was used for dark toxicity purposes. All dishes had undergone the same preliminary procedure, but after the drugs were removed from the dishes and fresh media was added, the dark toxicity dishes did not undergo irradiation. These dishes remained in the incubator throughout the remainder of the experiment.

Cell Cloning and Counting

Both sets of cells, the irradiated and the dark toxicity cells, were allowed to clone for approximately 3–5 days after the irradiation date, until they reached a size of approximately 1 mm in diameter, which allowed for their counting. The clones were examined periodically to inspect their size. Clones too small would be difficult to count. On the other hand, if the clones were allowed to divide too much, the cells may detach from the clone and produce their own colonies.

After the clones reached an appropriate size, they were set and stained. This procedure was performed for both the irradiated and the dark toxicity experiments. The old media was removed and washed twice with phosphate buffered saline solution (PBS). The clones were then fixed by the addition of methanol, which sat on the cells for about 5 minutes, and was then removed. The dishes dried and the clones were stained with crystal violet. The clones were counted with a colony counter, and both the irradiated and the dark toxicity clones were analyzed.

## RESULTS

The scope of the study was to compare the phototoxic properties of

Al phthalocyanine Tetrasulfonate produced by Ciba Geigy and Porphyrin Products. The photosensitizers produced were presumed to possess identical phototoxic behaviors. However, recent animal studies performed by our and another facility indicated differences in phototoxic behaviors between the two photosensitizers.

The photosensitizer manufactured by Ciba Geigy was received as a solution at a concentration of 300 ug/ml. The photosensitizer manufactured by Porphyrin Products was received as a powder, which was dissolved in .1 N NaOH and then diluted with Phosphate Buffered Saline (PBS) solution with calcium and magnesium. Porphyrin Products had substantially higher absorption peaks than Ciba Geigy. For instance, at the desired wavelength of 675 nm, Porphyrin Products had an absorbance of about 1.3, and Ciba Geigy had an absorbance of about .43. Since the Porphyrin Products' photosensitizer had the highest absorbance, it was diluted with PBS to lower its absorbance to that of the Ciba Geigy photosensitizer. The two photosensitizers were adjusted to have approximately the same absorption of approximately .43.

The intended number of cells for each petri dish was targeted at 250 cells/dish. Inability to obtain a constant number of cells for all the dishes may result in inaccurate data. Therefore, control dishes were plated out for every set of experiments performed. Table 2, containing data from the dark toxicity study, and Figure 1 below, containing data from the phototoxicity study, both present the control group at a drug concentration of 0 ug/ml, which reveals close consistency of cell numbers for every experimental set.

## Table 1
### Dark Toxicity of Higher Drug Concentrations

| Concentration of drug (ug/ml) | Number of surviving clones | |
|---|---|---|
| | Ciba Geigy | Porphyrin Products |
| Lowest 0 ug/ml | 207 | 179 |
| | 187 | 185 |
| | 193 | 184 |
| Highest 100 ug/ml | 185 | 183 |
| | 174 | 201 |
| | 183 | 172 |

Number of surviving clones at the control and highest drug concentration levels as a means of identifying cytotoxicity.

Due to the complex nature of the laser, frequent "inspections" were made throughout the irradiation period. Each experiment performed throughout the study utilized as many as 36 dishes, that were each irradiated for 3 minutes and 20 seconds; therefore, the laser was turned on for long periods of time at a high power. Periodic checks throughout the irradiation period were made to verify uniform power and thus to permit us to identify and correct a drop in power.

Following irradiation, the surviving cells were allowed to clone for several days. Periodic checks were made to identify the best day for fixing and staining the clones. If the clones were fixed and stained too soon, the clones would be too small to count. On the other hand, if the cells were allowed to clone for too many days, an excess of clones would hinder counting. For instance, cells belonging to a particular clone may detach from their original clone, attach to another part of the dish and create their own colonies. Therefore, the clones were counted when they reached a diameter of approximately 1 mm.

Dark toxicity studies were performed to determine whether or not there were any cytotoxic effects on the cell system without laser light. The data for the dark toxicity study carried out at high drug concentrations which ranged from the control, 0 ug/ml, to the highest drug concentration, 100 ug/ml, is shown in Table 1. The cells drugged with the Ciba Geigy and the Porphyrin Products photosensitizer at the highest concentration of 100 ug/ml as compared to the control show no cytotoxic effects. The data for the number of clone survival for the dark toxicity study carried out at the same time and at the same drug concentrations as the phototoxic study is shown in Table 2. In this case, the cells drugged with either photosensitizers reveal no cytotoxic effects for any of the drug concentrations when compared with the control. The number of surviving clones are fairly equivalent at all drug concentrations for both photosensitizer sources.

Cells were also irradiated with laser light tuned at a wavelength of 675 nm. The surviving clones were counted for each set of irradiation experiments. The data of the number of surviving clones after irradiation for the Ciba Geigy and the Porphyrin Products photosensitizers is shown on Figure 1. Three sets of experiments were carried out for each photosensitizer. The data identifies differences in the number of surviving clones between both photosensitizers at every drug concentration, when compared with the control, and reveals a higher cell killing for every set of experiments which utilized the Ciba Geigy photosensitizer. The average number of surviving clones of the irradiated and the dark toxicity experiments at each drug concentration is shown in Figure 2. The data shows no cytotoxicity effects at any drug concentration for neither photosensitizer. The data also shows there were no surviving cells for Ciba Geigy photosensitizer at a drug concentration of 1 ug/ml. All other concentrations show some cell killing, but for every drug concentration, the Ciba Geigy photosensitizer displays more cell killing than the Porphyrin Products photosensitizer. Figure 1 represents this data in form of a line graph. The control clone survival was fairly uniform for every drug concentration for both photosensitizers; therefore it may be presumed that the cell numbers for each dish were the equivalent, and that there was no sign of any dark toxicity. The curve representing the Ciba Geigy photosensitizer also drops much more rapidly at lower concentrations than the Porphyrin Products photosensitizer.

## Table 2
## Dark Toxicity Data

| Concentration of drug ug/ml) | Number of surviving clones | | | |
|---|---|---|---|---|
| | Ciba Geigy | | Porphyrin Products | |
| | Set #1 | Set #2 | Set #1 | Set #2 |
| 0 | 219 | 226 | 221 | 201 |
| | 202 | 200 | 216 | 164 |
| | 200 | 195 | 235 | 207 |
| .05 | 205 | 190 | 227 | 246 |
| | 201 | 186 | 209 | 220 |
| | 195 | 194 | 212 | 204 |
| .1 | 202 | 202 | 214 | 259 |
| | 229 | 210 | 205 | 251 |
| | 201 | 226 | 218 | 211 |
| .25 | 204 | 248 | 224 | 217 |
| | 211 | 232 | 223 | 177 |
| | 191 | 218 | 230 | 203 |
| .5 | 218 | 233 | 222 | 205 |
| | 219 | 244 | 227 | 224 |
| | 187 | 202 | 213 | 202 |
| 1.0 | 213 | 250 | 200 | 202 |
| | 199 | 217 | 193 | 201 |
| | 194 | 203 | 180 | 211 |

Study carried out in conjunction with the first two sets of irradiations at identical drug concentrations to identify any cytotoxicity. The total number of surviving clones at different drug concentrations is indicated.

The percent clone survival for each photosensitizer was determined at the different concentrations as shown in Figure 2.

The graph indicates there were higher clone survival rates for the Porphyrin Products than the Ciba Geigy photosensitizer for every drug concentration. At the drug concentrations of .05 ug/ml, .1 ug/ml, .25 ug/ml, .5 ug/ml, and 1.0 ug/ml, there was, respectively, a 90.1%, 35%, 5%, .5%, and 0%, clone survival for the Ciba Geigy photosensitizer. On the

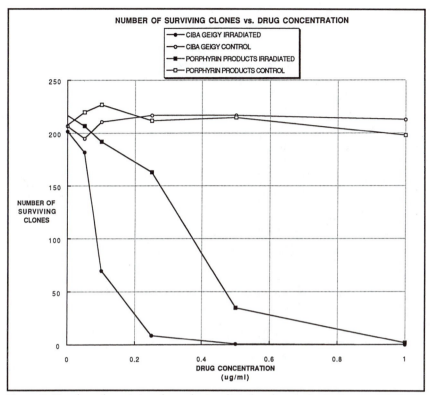

Figure 1: Number of surviving clones for irradiated and control sets at represented drug concentrations.

other hand, at the same drug concentrations of .05 ug/ml, .1 ug/ml, .25 ug/ml, .5 ug/ml, and 1.0 ug/ml, there was, respectively, a 95.4%, 88%, 75%, 17%, and 1% clone survival for the Porphyrin Products photosensitizer.

## DISCUSSION

The photosensitizers produced by Ciba Geigy and Porphyrin Products were used in animal studies conducted by our and another facility. The two photosensitizers appeared to show different phototoxic outcomes, and differences in the drug concentration needed to obtain the same clinical response. Twice the concentration of Porphyrin Products photosensitizer was needed in order to obtain equivalent clinical responses over the concentration of the Ciba Geigy photosensitizer.

Spectra created from the two photosensitizers indicated higher absorption peaks for the Porphyrin Products photosensitizer at equivalent drug concentrations. Adjustment of the Porphyrin Products photosensitizer allowed for a comparative phototoxic analysis between the two photosensitizers.

Distribution of cells was extremely accurate. For the different sets of experiments performed for the dark toxicity and the phototoxicity study,

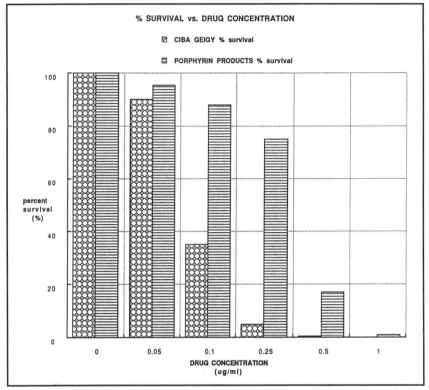

Figure 2: Percent clone survival at diffrent drug concentrations utilizing Al phthalocy-anine tetrasulfonate manufactured by indicated sources.

fairly equivalent cell numbers were obtained and, as a result, provided fairly accurate and comparable data.

Since uniform power of the laser had to be maintained throughout the irradiation period, frequent inspections were made to identify a drop in power. Periodic observations successfully maintained a power of 1414 mW.

After the clones were allowed to obtain a reasonable size, approximately 1 mm, the clones were counted with a colony counter. If the clones were too small, then they were not counted. However, if the clones were larger than the average size, they were still counted. This process was accomplished for every petri dish plated, which include the irradiated and dark toxicity clones.

Dark toxicity tests performed for the Ciba Geigy and Porphyrin Products' photosensitizers displayed no cell death. This observation was seen from the dark toxicity study performed in conjunction with the phototoxic experiments, which included drug concentrations at 0 ug/ml, .05 ug/ml, .1 ug/ml, .25 ug/ml, .5 ug/ml, and 1.0 ug/ml. The other dark toxicity study carried out at higher drug concentrations, ranging from 0 ug/ml to 100 ug/ml also displayed no cell death. This signified that there were no toxic effects for either drug at any of these drug concentrations.

The number of surviving clones as compared to the drug concentration indicated there was higher number of clone survival for the cells drugged with the Porphyrin Products than the Ciba Cadge photosensitizer. It would be considered safe to speculate that the photosensitizer produced by Ciba Geigy is slightly over twice as potent as the photosensitizer produced by Porphyrin Products.

## Literature Cited

1. Boyle, D. G., and W. R. Potter. "Photobleaching of Porphyrin II as a Means of Eliminating Skin Sensitivity." *Photochem. Photobiol.* 46 (1987): 997–1001.
2. Dahlman, A., A. G. Wile, R. G. Burns, G. R. Mason, F. M. Johnson, and M. W. Berns. "Laser Photoradiation therapy of Cancer." *Cancer Res.* 43 (1983): 430–34.
3. Dougherty, T. J. "Photosensitizers: Therapy and Detection of Malignant Tumors." *Photochem. Photobiol.* 45 (1987): 879–89.
4. Dougherty, T. J. "Photodynamic Therapy (PDT) of Malignant Tumors." *CRC Critical Reviews in Oncology/Hematology* 2, No. 2 (1984): 83–116.
5. Dougherty T. J., A. G. Wile, R. G. Burns, G. R. Mason, F. M. Johnson, and M. W. Berns. "Laser Photoradiation Therapy of Cancer. *Cancer Res.* 43 (1983): 430–34.
6. Dougherty, T. J. "Activated Dyes as Antitumor Agents." *J. Nat'l. Cancer Inst.* 52 (1974): 1333–36.
7. Gal D., "Effect of Photosensitizers in Chemical and Biological Processes: The MTO Mechanism in Photodynamic Therapy." *Biochemical and Biophysical Research Communications* 186, No. 2 (31 July 1992): 1032–36 (UI: 92359979)
8. Gomer, C. J., D. R. Doiron, N. Rucker, N. J. Razum, and S. W. Fountain. "Action Spectrum (620–640nm) for Hematoporphyrin Derivative Induced Cell Killing." *Photochem. Photobiol.* 39 (1984): 365–68.
9. Hampton, J. A., and S. H. Selman. "Mechanisms of Cell Killing in Photodynamic Therapy Using a Novel In Vivo Drug/In Vitro Light Culture System." *Photochemistry and Photobiology* 56, No. 2 (August 1992): 235–43. (UI: 92366659)
10. Kimel S., B. J. Tromberg, W. G. Roberts, and M. W. Berns. "Singlet Oxygen Generation of Porphyrin Chlorins, and Phtalocyanines." *Photochemistry and Photobiology* 50, No. 2 (1989): 175–83.
11. Li, J. H., Z. H. Huo, M. L. Jin, F. Y. Zhao, W. M. Cai, M. L. Gao, M. Y. Shu, and J. Zou. "Photodynamic Therapy in the Treatment of Malignant Tumours: An Analysis of 540 Cases." *Journal of Photochemistry and Photobiology, B: Biology* 6, Nos. 1–2 (June 1990): 149–55. (UI: 91038595)
12. Mang T. S., T. J. Dougherty, W. R. Potter, D. G. Boyle, S. Somer, and J. Moan (1987). "Photobleaching of Porphyrins Used in PDT and Implications for Therapy," *Photochem. Photobiol.* 46 (1987): 853–58.

13. McCaughan, L. "Lasers in Photodynamic Therapy." *Nursing Clinics of North America* 25, No. 3 (1990): 725–38. (UI: 90341073)

14. Potter W. R., T. J. Mang, and T. J. Dougherty. "The Theory of PDT Dosimetry: Consequences of Photodestruction of Sensitizer." *Photochem. Photobiol.* 46 (1987): 97–101.

15. Roberts, W. Gregory, and Michael W. Berns. "Cell Biology and Photochemistry of Photodynamic Sensitizers." *SPIE* 1065; "Photodynamic Therapy: Mechanisms, 19–20 January 1989, Los Angeles, California," 175–81.

16. Roberts, W. Gregory, and Michael W. Berns. "In Vitro Photosensitization." *Lasers in Surgery and Medicine* 9: 90–101.

17. Roberts, W. G., K. M. Smith, J. L. McCullough, and M. W. Berns. "Skin Photosensitivity and Photodestruction of Several Potential Photodynamic Sensitizers." *Photochemistry and Photobiology* 49, No. 4 (April 1989): 431–38. (UI: 89265100)

18. Savitsky, A. P., L. V. Lopatin, N. A. Golubeva, Poroshina M. Yu, E. B. Chernyaeva, N. V. Stepanova, L. I. Solovieva, and E. A. Lukyanets. "pH Dependence of Fluorescence and Absorbance Spectra of Free Sulphonated Aluminum Phthalocyanine and Its Conjugate with Monoclonal Antibodies." *Journal of Photochemistry and Photobiology., B: Biology* 13, Nos. 3–4 (15 May 1992): 327–33. (UI: 92373423)

19. Traganos, F., D. P. Everson, C. L. Staiano, Z. Darzykiewicz, and M. R. Melamed. "Action of dihydroxyanthraquinone on cell progression and survival of a variety of cultured mammalian cells," *Cancer Res.* 40 (1980): 671–81.

20. Viherkoski, E. "Lasers in Medicine." *Annales Chirurgiae et Gynaecologiae* 79, No. 4 (1990): 176–81. (UI: 91345285)

21. Waldow, S. M., R. V. Lobraico, I. K. Kohler, S. Wallk, and H. T. Fritts. "Photodynamic Therapy for Treatment of Malignant Cutaneous Lesions." *Lasers Surg. Med.* 7 (1987): 451–56.

22. Weishaupt K. R., C. J. Gomer, and T. J. Dougherty. "Identification of Singlet Oxygen as the Cytotoxic Agent in Photo Inactivation of a Marine Tumor." *Cancer Res.* 36 (1976): 2326–29.

## References

Day, Robert A. *How To Write and Publish a Scientific Paper.* ISI Press, 1979.

## Notes

1.   Instructions for Contributors, *Plant Physiology* 101 (1993): i.

# 7

## Using Graphics in Scientific Writing[*]

### Definition and Purpose of Graphics

The term *graphics* refers to pictorial illustrations designed to accompany a text. Often such illustrations consist of words, numbers, and pictorial elements together. *Graphics* include a broad range of illustrations, from the intricately drawn and richly illuminated capital letters on medieval manuscripts to the ubiquitous "smiley face" and its variations, all designed to lend specific tonal or emotional qualities to the texts they accompany. In contrast, science writers restrict their use of graphics to the objective, quantified representation of significant data upon which experiments and findings are based. Science graphics help the reader visualize and therefore better understand the significant evidence for a text's claims. In certain cases, such as a photograph of a previously unknown organism or structure, graphics can also provide a form of documentation.

Of the many sorts of pictorial illustrations which may accompany texts, science writing relies primarily on these particular graphics: charts, tables, diagrams, line drawings, and photographs. These graphics share a common characteristic that serves the presentation of scientific information very well: They can be designed to maximize the amount of precise, detailed, complex information a reader can understand but do so in a relatively small space and in a relatively short time. In other words, science graphics are essentially *economical*, a desirable feature of science writing. Ideally, science graphics should be simply designed so they are accessible to the reader, but at the same time, they should offer the reader complexity of detail.

However, it is important to keep in mind that whatever their level of detail, graphics *support* a text; they are not an argument or an explanation

---

[*]  Patricia Hartz of the University of California, Irvine, wrote this chapter.

in and of themselves. Therefore, the writer must take care to *interpret and explain the significance of the graphic within the text*, at the place where the writer refers to the graphic. Furthermore, graphics alone should not be used to explain, evaluate, interpret, refute or review data—the text itself should fulfill those functions. Graphically depicted data provide a text with evidence, but it is still up to the writer to explain its significance to the reader.

## General Principles of Graphic Use and Design

Science writers select a graphic design based on their specific purpose, but there are some general principles of good graphic use that science writers should follow to ensure the quality and accuracy of the graphics that accompany their texts. Regardless of the specific kind of graphic a writer selects, the following principles should guide its design.

1. Graphics impose upon a reader's time, so they should always serve a clear function and should never be introduced gratuitously into a text. However, they should be included when the writer needs to convey complex information and when patterns and trends of data provide significant evidence for a claim.

2. Graphics should present data clearly and concisely, show complex relationships accurately, and provide the reader with an appropriately labeled, coherent picture of the data set. Remember that a graphic tells a visual story about the data, and like any other story, it must be clear and coherent in order to be understood, and all the characters must be given a name.

3. Graphic elements and design should focus the reader's attention on the values, patterns, and trends of the *data* rather than on the *design* of the graphic. Writers should avoid content-free decoration such as frames or borders, unnecessary symbols, and the use of color, shading, or other technical effects which do not correspond to or represent actual data. Graphics should demonstrate *the significant variation of data*, not unnecessary elaboration of design.

4. Graphic design must never distort or falsify the data. The following two rules should be strictly followed in designing a graphic:

a) The number of variable dimensions represented in the graphic should not exceed the number of dimensions in the data.

b) The representation of numbers on the graphic should be directly proportional to the numerical qualities represented, ensuring that the graphic gives an accurate representation of scale.

5. Graphics must be placed in the text so that they are closely integrated with the statistical and/or verbal descriptions of the data set they represent. Data should not be quoted or referred to out of context, forcing the reader to riffle through pages trying to find the supporting graphic.

## Manuscript Conventions for Graphics

Text and graphic should be run together whenever possible, but graphics should not be set off with unnecessary lines or other intrusive or distracting marks. When possible, graphics should be laid out horizontally, slightly greater in length than in height, at a proportion of approximately 1:1½. These proportions are both aesthetically elegant and encourage the reader to focus on the data in the same manner that readers generally read, from left to right. Unless the pattern of the data itself makes another shape desirable, the standard proportions should be used.

Graphics large enough to take up their own page should fit within the standard 1" margin, just as the text does. Oversized graphics should be neatly folded to fit the size of the text and appended to the end of the text, and the location of the appendix should be indicated in the relevant paragraph in the text. Graphics not central to the text but of more general information should be placed at the end of the text in a separate appendix and given a descriptive title.

All graphics require descriptive, economical but complete titles, and clear, detailed, thorough labeling. The labels should be written on a graphic from left to right whenever possible and preferably in straight lines; angled labels should be avoided because they are difficult to read. Title and labeling words should be written out in full except for standard abbreviations (psi or cm, for example); other codes and symbols should be avoided. Science writers do not ordinarily use graphics that require legends. Instead, they design graphs so that all values are represented directly on the graphic itself. In other words, a graphic should be appropriately designed and sufficiently labeled so that no legend is needed. However, important events or values may and in many cases should be annotated or identified on the graphic itself. Explanatory information can also be included in notes at the bottom of the graphic.

The same typeface used for the text should be used for its graphics, and the typeface should have serifs. Though block letters may appeal to the lay person as appearing more "scientific," in fact, the block style is neither intrinsically nor conventionally scientific, and perceptual studies suggest that readers have greater difficulty distinguishing block letters than those with serifs. For a similar reason, the title, labels, and abbreviations, or any words on the graphic should have *both* capitalized and lowercase letters. STUDIES SUGGEST THAT WORDS WITH ALL CAPITALS ARE MORE DIFFICULT TO READ, AND THEY ALSO GIVE THE EFFECT OF YELLING AT THE READER. In contrast, using type with serifs, and using the standard rule of capitalizing only the first letter of title words (except for articles and prepositions), give readers the greatest ease in character recognition.

As another convenience for the reader, graphics should be numbered consecutively in a text, unless the text contains only one graphic, in which case no number is needed. If the text contains more than one graphic, and those graphics are a mixture of types (a table, a chart, and a diagram, for example), refer to the graphics as "Figure 1," Figure 2," etc. The figure number, followed by a colon, precedes the title of the graphic.

**Exercise 1:** Go to the library and examine a copy of *Science* and a copy of *Scientific American*. Focus specifically on the sorts of graphics accompanying the texts. What differences in graphics between the two publications do you notice? Write a paragraph in which you identify the differences and explain them based on the audiences they address. (Hint: Refer to Chapter 3). Which addresses the more general audience? What makes you think so?

**Exercise 2:** Assume the following graphic (below, p. 117) is the only one accompanying a text. Examine it closely. How many principles of good graphic use does it violate? List as many as you can. How does each violation interfere with the reader's understanding?

### Integrating Graphics and Texts

It is absolutely essential that the science writer understand that the functions of graphics and the written text are different and hierarchical: Graphics *serve as evidence* for the written text. They *support the written text*, but they do not constitute an argument in and of themselves. In order for evidence to be meaningful to the reader, its *significance to the writer's discussion or argument must be made clear.* Therefore, the writer must integrate references to graphics into the written text so that the reader (1) can find them quickly and easily, (2) has a description of the experiment on which the data is based, (3) has the relevant pattern or trend of the graphic identified and explained, and (4) is given the conclusions suggested by the data.

The reader can find the correct graphic easily if the graphic is clearly numbered and near the paragraph referring to it. The reference should be placed in parentheses. For example, "(See Table 9)." The graphic should not be referred to until the experiment or event on which the data is based has been described. Then the significant pattern that the data represents should be explained. Immediately following this, the graphic should be referred to in parentheses. But the logic which connects the graphic and writer's conclusion based on the data must be given to the reader, since the data cannot argue itself.

### Titling Graphics

Titles for graphics in science writing are not optional; they are important pieces of information your reader needs in order to understand the significance of graphical data for your discussion. Graphics which are not identi-

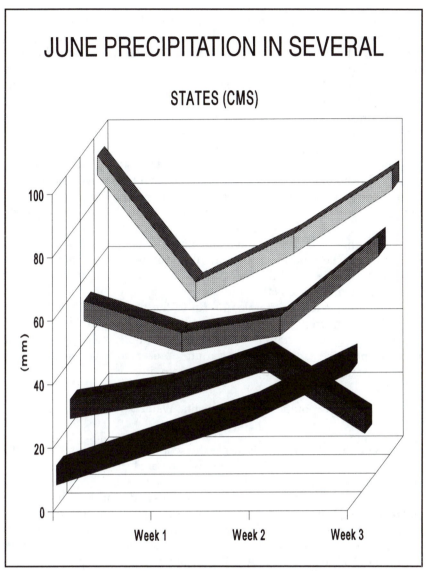

JUNE PRECIPITATION IN SEVERAL STATES (CMS)

fied by a title are more likely to confuse the issue for your reader than they are to clarify it. Well-selected titles contribute to the reader's understanding when they uniquely distinguish *each* graphic by describing the data represented.

Describing the graphic in a title is not as easy as it sounds. Beginning writers are likely to confuse a descriptive title with a conclusion drawn from the data. For example, say you experiment with the effect of hours of sunlight on the growth of a plant. You gather your findings and create a chart that represents them. The chart shows the growth of the plants in relation to the number of hours of sunlight the plants were exposed to at different times. The data suggests that the effect of decreasing sunlight is to slow down the plants' growth. What would a good title for such a chart be? It would *not* be "Decreased Sunlight Retards Plant Growth." Why? Because this title does not describe or identify the graphic data at all—it draws a conclusion from the represented data, and such *conclusions are part of your discussion in the text, not part of the graphic.* Avoid this error when selecting graphic titles.

Good titles have several other characteristics as well. To say they uniquely identify each graphic is to ensure that the distinctive feature of the graphic is the main part of the title. For example, assume you have created a graphic using data derived from an experiment to discover the effect of water salinity on the rate of evaporation. An appropriate title would be, "Evaporation Rates for Saline Water Solutions." Note that this title correctly uses nouns as its principal elements; this is because the function of nouns and noun phrases is precisely to name things. Other parts of speech, such as relative clauses and participles should be avoided. For example, these are not good titles: "Saline Water Solutions that Evaporate at Different Rates" or "Evaporating Water Solutions."

Sometimes subtitles are also useful to the reader. Subtitles occur in parentheses centered beneath the main title, and they should be restricted to secondary but important information. For example, if a condition of the experiment is the same for every datum represented, one could indicate that in parentheses centered beneath the main title:

Evaporation Rates for Saline Water Solutions
(T=40°C)

Or one might choose to indicate the scope of the data—for example, "(by year)." Or the unit of measurement or analysis, "(in cm)." Or the number of test subjects, "(number of test subjects=500)."

Some general principles apply to both titles and subtitles. They should not repeat or reproduce discussion from the accompanying text, nor should they make extended comment; they should not provide irrelevant information; they should not express the writer's subjective feelings; and they should not aspire to or actually achieve humor.

Both titles and subtitles use the standard rules of capitalization: the first word is always capitalized, and all following words except for articles and prepositions. Common abbreviations for units of measurement may be used in subtitles, and need not be capitalized if they are not ordinarily capitalized: for example, "(psi)." Titles should be centered above the graphic. However, long titles are generally left-justified if they are longer than three lines.

**Exercise 1:** Devise a title appropriate for a graphic demonstrating the effect of length of sunlight on plants. Then devise an appropriate subtitle.

**Exercise 2:** Evaluate the following titles:

a)          Data Disproving Dr. Smith's Ludicrous Theory

b)          Occurrence of Intestinal Parasites in Domestic Cats
(Suggesting Monthly Inspection of Fecal Material by Owners)

c)          Distribution of Iridium on the Yucatan Peninsula
(Field Trip Financed by National Science Foundation Grant)

## Selecting a Specific Graphic

The first question the writer should ask is whether a graphic is necessary at all. Most *simple qualitative* relationships can be described in the text. For example, a structural engineer's description of the deterioration of steel span bridges within a county could be presented clearly without using a graphic. However, *complex qualitative* relationships might very well be better expressed in a diagram. For example, if the structural engineer wanted to represent the complexity of causes that affected the bridges being analyzed, causes such as overload due to traffic congestion at high-use periods, improper truck routing, and periodic river flooding, the engineer might very well devise a diagram such as a flow chart to clarify causal relationships for the reader.

But what if the engineer wanted to represent the number of deficient bridges found in every county in the state? When the writer wants to express *exact quantitative values*, as would be the case here, a table is the most useful graphic choice. Suppose our engineer also wanted to represent the estimated rate of deterioration on the state's worst bridges. In that case, when *quantitative patterns or processes* must be represented, a chart or graph is most useful.

If we imagine that the engineer also wants to show exactly which of the weight-bearing beams on the worst bridges are in immediate danger of collapsing because of stress faults, it is easy to see that line drawings would be the best graphic choice. Since a line drawing is a schematic image of an object that emphasizes a specific part, the engineer could show the area of deterioration clearly by eliminating the irrelevant parts of the bridge from the drawing and by making the danger zone or fracture site itself the focus. Alternatively, if the engineer discovers that rust and corrosion threaten specific bridge supports, photographs of the sites, with a scale included to allow the reader to estimate the amount of metal actually eaten away, would provide the reader with the clearest understanding of the situation.

Not only are graphics selected on the basis of the writer's purpose, but also on the basis of the nature of the relationships depicted. Most science graphics are either based on chronology (for example, a time-series or a time-space series), or they show relations between variables. The major difference between these two types is that a time-series graphic need not necessarily suggest a causal relationship between a variable and the passage of time, but may simply *describe* a process instead. Relational graphics, however, those which show the relationship between two or more variables, frequently suggest an "X causes Y" proposition. A writer needs to consider carefully the sorts of relations being depicted to make sure that chronological relationships are not graphically expressed as causal, and vice versa, that causal relationships are not graphically expressed as mere chronological changes. In general, unless an instructor asks the student writer for a specific graphic form, or unless the writer's discipline uses specific, conventionalized graphic forms (as in computer science, for example), the writer should select a graphic type on both the basis of the writer's purpose and the nature of the data.

### Individual Kinds of Graphics

Tables: A table is a collection of individual pieces of information arranged in rows and columns according to category and type. Use a table to show exact numerical values and to allow the reader to make many detailed, discrete comparisons. Tables can be arranged chronologically, like a train schedule, or they can be used to represent items in order from whole to part or values from high to low. Tables are useful to the reader only if the rows and columns are designed logically and if the design allows the reader to locate the information quickly. Poorly designed and/or cluttered tables are

### Parts of a Table

| Sub Head | Column Spanner | | | |
|---|---|---|---|---|
| | Column Head | Column Head | Column Head | Column Head |
| Row Head | Field Spanner | | | |
| | | | | |
| | | | | |
| Row Head | | | | |
| | Field Spanner | | | |
| | | | | |
| | | | | |
| | | | | |
| Explanatory Note: | | | | |

more confusing than helpful to the reader, so writers should avoid including too many different sorts of information in the same table. In general, the writer may vary information by column or by row, but not by both at the same time. As another general rule, tables are a good choice when the number of data sets is twenty or less. A chart or graph is better suited for larger numbers of data sets.

### Columns and Rows

Representing the data may not require all the parts that the table above has, so only as many categories as the data require should be used. Writers should create column headings to designate dependent variables and row headings to designate independent variables or categories. For example, a table showing the relationship between biodiversity and land area might list the names of species of plants as row headings and the land sampled as columns headings. Each table cell could then contain the number of individuals of each species in a specific land sample or plot.

In general, each column heading should apply to all the members in the column and each column category should be unique. Column headings are centered if they are less than one line; longer column headings are left-justified. Columns may also use subheadings, and these will frequently be units of measurement. Such subheadings should be centered between parentheses beneath the column heading. If specific columns are referred to in the text, those columns should be numbered using arabic numerals placed between parentheses beneath the column heading or subheading.

As with column headings, row headings should be unique to their row and apply to all members of their row. They should be brief and clear; and any needed secondary information should be placed in a note, as located above. Row headings are left-justified and use standard capitalization, just as columns do. The stub head designates the row headings, and should not be left blank unless the table's title clearly identifies the rows. For example, a stub heading for the biodiversity table described above could be "Species of Plants." However, if the table's title was "Species of Wild Plants in Suburban Green Belts," the stub heading could be omitted and the graphic made more economical.

One of the virtues of a table is its relative flexibility in adding or subtracting categories. In horizontally ruled tables, column spanners can be added to show relationships among a hierarchy of columns. For example, the biodiversity table could be designed to tell how many of each species were found in suburban *and* rural plots of the same size. In that case, a column spanner, such as "Sampled Area" could be used above the two column headings, "Suburban Green Belt" and "Rural Woodland," for example, indicating that "Sampled Area" is a broader category than the two column headings.

Field spanners can be used to designate a third variable, in which case the row heading for each new field spanner is repeated, as shown above. For example, the biodiversity table could represent the species of each category, "Suburban Green Belt" and "Rural Woodland," at different altitudes, those occurring only below one mile above sea level and those occurring only above it. Using field spanners would allow the writer to add this level or category of data. Obviously, a writer would not use only one field spanner in a table.

## Placement of Data

All the data in a table, whether text or numbers, should be aligned consistently for the ease of the reader. Text is generally left-justified, while whole numbers are generally right-justified. Formulas are aligned on the equal sign. If dates and times make up part of the table, they should be represented in a consistent format: for example, either "7/10/95" or "July 10, 1995," but not both, so that they can be aligned.

Spaces in the table for which there is no data should not be left blank; an absence of value is indicated by inserting ellipses points or by using a zero; the abbreviation *N/A* and the phrase *Not applicable* are not appropriate for science writing. Instead, explain the absence in a note. Nor do science writers use ditto marks to show repeated values in a column. And for the reader's convenience, if the table consists of long columns of numbers, a blank line can be inserted every five or ten lines.

## Table Notes

The most common types of notes a table might have are those which cite sources for the data in the table and those which explain or clarify the data in the table, such as degree of accuracy, or how values were manipulated or rounded off or how the data were collected. Source notes should always be listed first, before explanatory notes, and the source citation should be preceded by *"Source."*

## Numbering Tables

Number tables in the order they appear and cite them in the same order in the text, but do not number a table if it is the only one in the text. (Also, remember that if two or more tables are used, they should be of the same design, if possible, for a single piece of writing.)

**Exercise 3:** Starting with the various bits of information given for the "biodiversity table" examples above, and adding categories and data you invent as needed, prepare a table that has all the parts of the table depicted on p. 120.

## Charts

Charts are used for representing quantitative relationships, such as the values of numerical data, a process or change over time, or patterns of data. They are frequently used to predict future data based on data already known. Charts should be simple to apprehend, clear, and concise, and the design should not distort the data.
    Parts of a Chart:

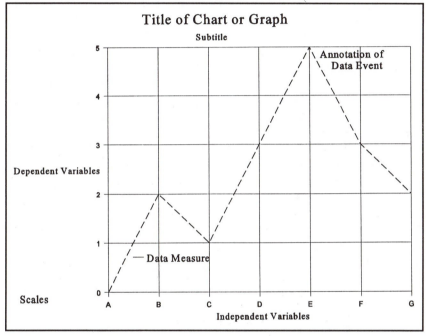

## Title and Scales

Every chart and graph should have a brief but adequately descriptive title, preceded by a figure number if needed. Scientific writers avoid humorous, partial, or obscure titles. Charts and graphs require at least two scales, one along each axis, and can be selected to show change linearly, logarithmically, or exponentially. The scale label must make clear which scale is being used.

## Data Measures

The graphical element that actually locates the data on the graphic is called the *data measure*. The data measure on charts can be represented by a variety of dots, lines, or areas. Data measures should be selected according to whether the writer needs to represent exact values or trends. A point or dot will indicate an exact value, whereas a line would indicate a trend. It is difficult for a reader to estimate the value of or to compare areas or volumes, particularly if they are irregular in shape, so these should be avoided if possible. Bar charts, however, which do compare areas, are frequently used by science writers, but their appearance, their size, shape, and density of shading, should be consistent with the data values they represent. Science writers do not ordinarily use pie charts, and it is best to stay away from circles and spheres in general, unless your instructor gives you other directions, or unless the nature of the data makes a circle or sphere especially desirable.

Data measures can also be represented by pictorial symbols, but science writers ordinarily avoid using them. The use of symbols should be restricted as much as possible to those symbols accepted as conventions in the area of science the writer is working in. If other symbols must be used, they should be simple geometric forms, circles, triangles, squares. Color should be used sparingly and only for emphasis. It should not be used as the only determinant of a value, nor should it be used to make critical distinctions. Also, avoid red when possible since it is the color most likely to prevent color-blind readers from reading the chart or graph accurately.

Charts and graphs are the graphic form most prone to abuse by technical effects that contribute nothing to the reader's understanding of the data, and at the same time distract the reader from the relevance of the data. Particularly if you have access to computer graphics programs, avoid elaborate shadowing techniques, extra borders, and annoying pattern selections, such as elaborate cross-hatching or stripes which seem to shimmer on the page or give other unpleasant and distracting optical effects, especially on bar graphs. Although most computers programs allow a variety of "creative" bar graph designs, grave restraint should be exercised in design selection. Try not to stack or overlap the bars on bar graphs; they should be placed next to each other. The goal is to make each value represented as visually distinct as possible and to reflect the data accurately. Therefore, elaborate pains should be taken to avoid a form of graphic representation that visually distorts or obscures the data's actual values.

### Gridlines and Frame

Charts and graphs are generally framed by the perimeter of the grid on which the data values are plotted, with the gridlines usually about half the weight of the frame lines. However, it is sometimes possible to delete the grid behind the represented values altogether if the trend or pattern of the data is more significant than actual values.

The upper and right hand frame may also be omitted if the data pattern does not extend into those areas. In other words, a writer can choose to

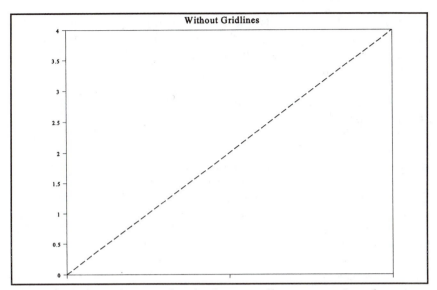

present only the scale range that the data actually represents in order to save space.

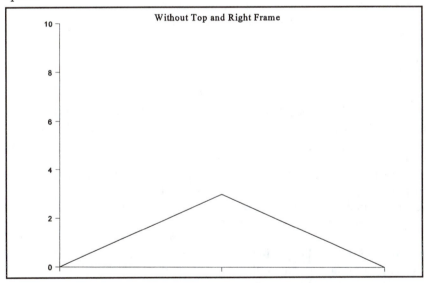

**Exercise 4:** Here is a formula that describes the classic Malthusian account of population growth, were it to be unlimited by food scarcity or ethical constraint: $x_{next}=rx$, where x represents the population for the first year in which the population is measured, and r represents the rate of population growth. For example, if the rate of population growth is 1.2, then if this year's population is 10, next year's will be 12. If this year's population is 15,000, next year's population will be 18,000.

Stipulate values for x and r, and determine the population for ten years. Then draw a graph, appropriately titled and labeled (create an appropriate title and labels), to represent your data. (Will your scale be linear, logarithmic, or exponential?)

**Exercise 5:** Here's an example of a more sophisticated formula to represent population growth: $x_{next}=rx(1 - x)$. It is a modification of the classic Malthusian case and a variation of what population biologists refer to as a "logistic difference equation." The formula includes a variable which represents a factor which could limit population growth, such as the natural death rate and/or the additional death rate from starvation and predation. This formula expresses the largest conceivable population of the represented species as 1, and 0 would represent extinction.

Using a starting population of .02 (meaning x=.02) and a value of 2.7 for r, generate the first thirty-five years' populations. Represent your data in a well-designed, titled, and labeled table. How will you make reading the values easier for your reader?

Now represent your data in a well-designed, titled, and labeled graph. Which form, table or graph, is more effective in showing the trend of the data? Which is more effective in showing exact values? Given your findings, do you think that the data are better expressed by a table or by a graph?

## Diagrams

Diagrams are enormously useful since they can be used to demonstrate a wide variety of relationships: spatial, such as a those represented on a map; sequential, such as those represented by flowcharts; and the relationships of parts to a whole, such as the visual aids to assembling a piece of lab equipment. They can also illustrate causal, connective, dependent, and compositional relationships. Students may also find diagrams an excellent means of organizing their material to assist their own learning as they record their daily work. Such diagrams are often the "first draft" of the graphic which will ultimately illustrate the text.

Some disciplines, such as computer science, use formal diagramming conventions particular to their field. If the discipline in which you are working uses its own conventions, your instructor will introduce you to them. But diagrams are routinely used in all fields of science, for both lab journals and formal papers. In lab situations, they are often used to demonstrate the assembly and/or use of an object or the relationship of parts of a process.

Diagrams are especially useful in clarifying complex relationships, as long as the design selection accommodates the structure that it represents. For example, an appropriate diagram design to represent the structure of the United States' chain of military command would be a tree form, which accommodates hierarchical relationships. A web diagram, however, would be appropriate to represent more complex relationships, such as chemical processes in the cell, for example.

Diagrams can be composed of square or rectangular grids of various dimensions and can be designed to represent data in two logical dimensions, much as a table does. Such linear processes can be diagrammed sequentially, with connections between diagram elements indicating movement, whether forward, backward, or both, if necessary, and indicated by arrowheads on the connecting lines or arrows alongside them, if this does not interfere with clarity.

Science diagrams should use either the conventional symbols of the writer's discipline or sub-discipline, or simple geometric shapes. Larger shapes should designate the more important or controlling objects in a diagram, and a diagram should not use too many different kinds of shapes. If more than half a dozen are necessary, the writer should probably give the information in two separate diagrams. The symbols should be labeled, within the shape if possible, and immediately outside it if not. The lines linking the symbols may be labeled as well, and if space permits, may be written on the line itself; if space does not, the label may be written just above it. Whenever possible, labels should be written horizontally for the reader's convenience. As with other graphics, diagrams must have titles and numbers if the text contains more than one diagram.

Whatever the overall shape and pattern of the diagram as a whole, all the diagram elements should be connected to at least one other element. None should be left standing alone. Diagrams should also eliminate all unnecessary elements of the process or structure being depicted in order to maximize the reader's ease of comprehension. If more than one diagram accompanies a text, they should be as similar as possible in their schematic elements: lines should have the same weight, boxes the same shape, etc.

Examples of possible diagram designs and elements:

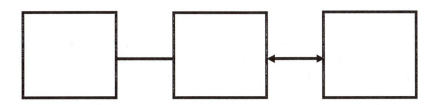

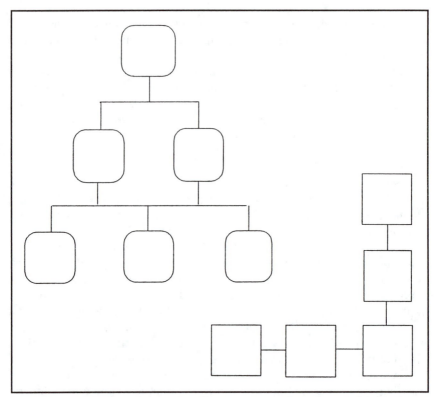

**Exercise 6:** After studying the diagram examples above, think of one process or structural relationship that each would adequately represent, adding or subtracting elements if needed. Add arrows if needed, and label the diagram elements.

**Exercise 7:** Examine the diagram on p. 156. Then, using different symbols for each form of contagion, create a diagram that demonstrates Girolamo Fracastoro's germ theory of disease (pp. 219–26).

## Line Drawing

Science writers use line drawings to show particular features or specific characteristics of a complex object. Often such drawings are designed to show the reader how something works. Line drawings can also accompany descriptions of how to assemble, install, use, or maintain something. In creating a line drawing, the writer must take the point of view of the reader in determining exactly what to illustrate. For example, if only the nucleus of the cell is the focus of the drawing, the other cell contents would not be represented. But if the cell wall is the object of focus, none of the contents of the cell would be represented.

Keep line drawings simple—they should be neat, clear, devoid of unnecessary detail, and they should be in scale, although sometimes a line drawing magnifies a specific feature if necessary for clarity. As an aid to accuracy, the writer should have access to the object to be drawn, or else a very good print, photo, or photocopied image. Drawing equipment such as rulers and protractors should be used when they can make the drawing neater and more accurate. Each illustration must be labeled clearly, using arrows where necessary to indicate assembly or direction. As with other graphics, if more than one line illustration is used, they should all have the same style, line weight, font, etc.—do not mix different kinds of illustrations.

**Exercise 8:** Examine and evaluate the line drawing of a pipet bulb on p. 141. Then prepare a drawing of a piece of equipment found in your lab.

Photographs: Photos are used to show readers exactly how something looks and sometimes to give evidence of authenticity and/or to document the existence of something. Photos should restrict themselves to the specific object or phenomenon as much as possible. If a specific part of the photographed object is the significant element, then whenever possible, the photograph should be taken from an angle that exposes the part most fully. If necessary, an additional photograph of the entire object can be included to allow the reader to visualize spatial relationships. Photographs which distort the object depicted should be rejected. A rule or other means of determining scale should be included in the photograph if size is of significance.

Photos should be restricted to only the necessary surrounding context; unnecessary context should be cropped from the photo. A photograph of a giant land snail ascending a rock should not have the feet of the researcher in the frame, for example.

Photographs intended to accompany a text should be of the highest quality possible, sharp, in focus, and well illuminated. It will often be useful to the reader if the quality of photos is sufficiently high to allow legible reproduced copies.

Photographs must be clearly labeled, and are sometimes accompanied by a brief identifying or explanatory text beside or beneath the photo, in addition to its caption. This text is called a *callout*. If arrows indicating significant features in the photo are necessary, a ruler should be used to draw a straight line between the callout and each significant feature.

**Exercise 9:** Evaluate the photograph of a cownose ray on p. 174 from the point of view of a scientist writing about the ray's reproductive processes. Then, using the following information, write a callout designed to accompany the photo, placing arrows between text and photo to connect specific features: Cownose rays have powerful tooth plates, composed of a series

# 8

~~~~~~~~~~ ⤫ ~~~~~~~~~~

# Description/Perspective

Description is the writing complement to observation; it *renders* or gives body to the observation. Descriptions create pictures for a reader through precise and objective language organized in a consistent pattern. *Picture* is not meant in its purely imitative function nor, technically, must language be limited to words. Mathematics may be considered as a kind of language, and mathematical descriptions can render a simulacrum of phenomena through such a language. More narrowly defined, description creates a representation of the object through the particular "lens" used to observe it. This chapter will concern itself with using words as the medium for making descriptions.

Descriptions report the attributes of an object, event, or phenomenon in order to render a representation. One need not KNOW what the event is to ascertain the attributes. Given sufficient attributes, it is possible to assign the unknown an identity. But even if identity is misassigned, it is possible for subsequent investigators to correct an assessment of identity if the rendering of the attributes is comprehensive.

## Audience and Purpose

Identify the audience and purpose of your description. Does your purpose call for a description of a static object and thus your description will be spatial? Do you need to describe something that takes place in time (rather than in space) and hence will require a temporal orientation? Should the reader be able to *recognize* the object your describe? To *differentiate* one form from a visual field of related forms? To *manipulate* the object you describe? To *repeat* the steps in a process? To give directions?

Assess how much *detail* is required to satisfy that purpose for your designated audience. "It's the yellow convertible" might be sufficient detail if you toss your car keys to your friend and point to the parking lot. But more detail, and a different kind of detail, is required to differentiate the short-term metabolic effects of Insulin-like Growth Factors (IGF) from the long term mitogenic effects in a scientific paper prepared for an expert audience.

# THE STRUCTURE OF DESCRIPTION

Sense perception is multifaceted and random, and many perceptions can be made simultaneously. A single visual image, for example, presents us with a complexity of sensory data. When we observe a flower, our impression of its form, color, and position among other flowers and our impression of the clarity of the image (whether we see it on a cloudy day or a clear day) occur at the same time. But a description of this perception, as rendered in the previous sentence, requires a linear order of words that breaks the perception into parts.

*We cannot write about simultaneous events simultaneously.*

Simultaneous events, even at the micro-level of perception, must be differentiated, partitioned, broken up into parts, in order for us to write about them.

A description in words must be linear because we read in a linear, temporal fashion, from left to right, with one word following another. To transfer perceptions into writing, the writer must impose form and order on perceptions. Description renders perception through writing.

The order of description follows the logic of writing.

- In writing, one usually moves from general to particular.
- In description, the pattern moves from *whole to parts*.

The ordering of description is important because it makes detail coherent. The following examples will demonstrate this point.

## Examples

In the following three descriptions, the writers describe the same object. Read through each of them and compare them generally. Which description offers the clearest picture of the object? Try to identify the strengths and weaknesses of each description. An analysis of their relative merit follows the last description.

### Description 1

The object in general has the overall shape of a flat, five-armed star. All the visible body components are jagged and rough on both sides, giving it the texture of rough sandpaper. It has an orange color at the top, and white and yellow on the bottom. The object is thickest at the center (about 2 cm), and the thickness decreases gradually to the tip of the arm whose thickness is about 1 cm. At the center, it has an imaginary surface of a circle about 4 cm in diameter. It weighs about 50 grams. The object has a five-fold star shape with roughly bilateral symmetry. There is a dorsal and ventral side but no noticeable anterior or posterior sides due to its symmetry. The color on the dorsal side is light orange with light orange yellow and dark red spots. In between the light yellow spheres are dots that are the size of sugar crystals which cover the whole dorsal side. The center of the object has a hole on the ventral side which is half a centimeter and half a centimeter deep. The color and structure on the ventral side is the same as the ventral side of the arms. All appendages are equal in size but different in shape. One is curved to the left and has the shape of shark's dorsal fin while the other bends approximately one and one-half centimeters up from the middle of the arm.

## Description 2

The shape of the object is semi-planar five-point star which is composed of a general body mass with five extensions radiating outward from the center like bicycle spokes. An imaginary line drawn connecting the tips of the spokes (like a tire) would measure about 10" in circumference.

The star-shaped creature has a form like a hand which rests palm-down. The undersurface of this star coincides with the palm of a hand, and the "back" side corresponds to the back of the hand. While the arms radiate from the center with radial symmetry, an axis of bi-lateral symmetry could be defined if the creature were cut lengthwise in half, from tip to tip.

While the dark orange "back" side of the star reveals a consistently jagged but unbroken surface, the underside (the palm) is different. A groove runs from the tip of each "finger" to a central opening in the "palm" measuring one-half centimeter in diameter and extendone one and one-half centimeters into the body mass.

## Description 3

The object in front of me originated from the sea. The sea is defined to be a body of salt water that covers a majority of the earth's surface. Within the sea are many forms of life, and they live a variety of ways. This one is a starfish, which belongs to the class of echinoderms.

Description 3 would have identified the object for you if you had not been able to form an identification.

Description 1 and 2 contain roughly the same data about shape, dimensions, and physical attributes. Both descriptions attempt to accommodate two kinds of symmetry: radial symmetry and bilateral symmetry. Description 1, however, lacks an overall pattern of organization. While groups of descriptive detail can be discerned—two related sentences describing color, two sentences dealing with appendages—those groups of detail are not integrated within a larger organizational scheme.

Because the writer shifts perspective on the object—alternating details from an "above" view, from a "below" view, and from a side view—the discrete, discontinuous points fail to add up to a unified picture of the whole. The orientation to the object provides the basis of an organization to the writing; multiple, simultaneous perspectives are reflected in the need for patterned organization. While it is appropriate to employ multiple perspectives in description, they should not be employed simultaneously lest they render the effect of looking through a fly's multi-faceted eyes.

Description 1 also modifies detail throughout the passage: Initially told the object is flat, we later learn it is 2 cm thick at the center. While the author describes color on the ventral side, gradations of color are not otherwise ascribed to a pattern of dispersal. Minor details (color) receive detailed treatment while more important information (picture of the whole) receives truncated treatment. For instance, a description of two curved arms is given, but no indication specifies which arms curve, an important consideration should flexing be a function restricted to few, not all, of the limbs. While the analogies (the surface of the creature is like sandpaper, the bend in the arm looks like a shark's dorsal fin) are effective for these parts, a coherent picture of a single object does not emerge.

Description 2 uses two images to organize the description of detail. The analogy of the bicycle spokes gives a picture of radial symmetry while the analogy of the hand accommodates the bilateral symmetry. Particular details are given in the context of their relevance to either one or the other kind of symmetry. Quantitatively, Description 2 gives fewer specifics, but the details provided are coherently structured. *Organized detail is more meaningful than random detail.*

Description 3 is not a description at all; it is a definition. It assumes knowledge about the creature already. While Descriptions 1 and 2 attempt to explain *what it looks like*, Description 3 tells you *what it is*. A definition implicitly contains an interpretation of the data; a description *renders* the data.

In selecting and reporting details, the writer may be caught between the need for relevance and the need for comprehensiveness. Description 1 attempted to be comprehensive, but the significance of the detail was lost by its scattered rendering. It is possible for description to be both comprehensive and structured; these desired attributes are achieved by the organization of details through the writing. Discrete dissociated detail—uncontextualized detail—is meaningless.

## Guidelines for Writing a Description

### 1. Describe the obvious

Make sure you are thoroughly familiar with your object, but do not take its identity for granted — this is a paradox. Do not identify an item as a frogfish and assume that your reader will know what such a fish is. The name identifies the object, and while identification offers a useful starting point, it does not create a picture. For instance, the Reading Selections section at the back of the book contains an extract from a treatise on epidemics by Hippocrates, a Greek physician (430 B.C.). Although Hippocrates identifies a disease as *causus*, the details he provides prevent a contemporary researcher from making a definitive identification of the disease. In contrast, Thucydides, in "The Plague of Athens" (also in the Readings), fails to identify the plague by name but his detailed descriptions are thorough enough for a modern physician to make a reasonable inference about the nature of the diseases. Because description creates a full picture, identification is not alone sufficient.

### 2. Use precise, objective language

> That big ugly virus I saw yesterday . . .
> The larval form of that moth was gross.
> Predation is depressing.

These examples contain subjective language—*ugly, gross, depressing*—that conveys more information about individual opinion than about the topic. "That big" virus gives general descriptive words but fails to specify size: one micron, half a micron? How big? Give specifics. Use standard units of measurement.

Specify the parts for items which have parts, (what the separate parts look like, how they fit together). Be specific about size shape, color, material, color, substance, scale, dimensions (weight, height, depth, surface area, etc.), orientation of parts, orientation of your description (top to bottom, inside to outside, front to back).

For spatial descriptions, rely upon specific words for form. Even if you do not have a clue as to what you have in front of you, grossly indicate the form. From one perspective, the moon appears as a disk; from another, a sphere. Consider your perspective in determining shape and use orientation words that relate to the object. A sea urchin is spherical not circular.

Where appropriate, use words like *square, elliptical, circular, round, spherical, disc-shaped, conical, helical, spiral, stellate, serrate, chiral, achiral, symmetrical, asymmetrical, continuous, discontinuous.* If you do not know a word in this list, look it up.

Be careful in selecting the language of orientation. For example, if describing a biological organism with a spherical shape (like a sea urchin), top and bottom are problematic terms. The sea urchin lives with its mouth to the rock and its anus presented to the sea. In spatial terms, the *top* of the creature is the physiological *bottom* of its digestive system; *dorsal* (back side) and *ventral* (belly side) apply only to bilaterally symmetrical animals. Similarly, for radially symmetrical animals, terms like *left* and *right* are meaningless because no location on the animal corresponds with these words, although these words do describe the perspective of a bilaterally symmetrical observer. Assuming knowledge of the mouth and anus, one could speak of an oral/aboral axis of symmetry. Words derived from bilateral symmetry are meaningless in cases of radial symmetry; similarly, the words for two-dimensional geometrical forms subvert descriptions for three-dimensional objects.

*Orientation is not simply spatial; it is categorical.*


### 3. Use a systematic pattern of organization

Nothing is too complicated to describe. Move from whole to parts. Break the whole into parts. Then describe the parts.

Among the ways to organize your description of parts are the following three:

If you describe an object in space, use (A) a *spatial sequence.* A description from top to bottom would be appropriate to describe a skeleton. For a creature with bilateral symmetry, a description might be framed with respect to that axis, starting with proximate features and moving to remote features. A car engine could be described from outside to inside or from inside to outside.

(B) Description of an object in terms of the way that *parts come together* is another option in organizing the description. An engineer would speak of an order of assembly; a biologist would speak a pattern of development or growth. This description could be ordered from the first step through consecutive steps, or from the last step to the first. Regardless of the specific application of this concept, the pattern must be chronologically consistent.

For organic creatures, you may write about the order of assembly from the standpoint of developmental biology in terms of which structures are formed first in the development of a growing organism.

If the way that an object works is central to the description, organize your description in terms of the functional sequence.

(C) An organization of parts as they *function* requires consistent sequencing. The starting point for this pattern is use or operation—how something works or functions—rather than how the parts come together.

What do you do first, then second, then third? A description of a bicycle emphasizing its functional sequence (the way that one uses it) may start with a description of the whole, then move to the parts: the pedal, where pressure is applied; the gears and chain, which distribute the energy received from the pedal; then the wheels, which turn in response to the gears; and then the brakes, which stop the wheels. Notice that this order describes the functional sequence of propulsion; another sequence must be described for the whole sequence to be complete, and that would be the functional sequence in steering, an operation that occurs simultaneously with that of propulsion.

A description of the functional parts of an organism would be shaped by the needs of your discipline, whether, for instance, you were approaching this from the perspective of evolutionary biology or developmental biology.

Note: If you begin a description of an object based on the way the parts come together, do not switch to describing the way the whole works (unless the two otherwise go together). Do not switch your point of orientation until you have completed a full discussion of the one underway.

## Identifying Attributes

One may describe a thing or an event by identifying its attributes. The following questions will guide you in thinking about the important data that goes into a description. These questions are meant to spark your thinking. Simply answering the questions will not fulfill the function of description because description must be organized coherently.

### 1. What is it?

Identify the object. Is it solid, liquid, or gas? Is it animal, vegetable, or mineral? Find some category as the basis of classification so that your reader will have a general sense of what it is. Every thing, phenomenon, or event possesses a resemblance, even if purely heuristic, to another, and therefore has a class. Even if you encountered something totally unknown, like the alien monolith found on the moon in *2001: A Space Odyssey*, you can still offer tentative identification. If your identification is incorrect—if, for instance, you identified that monolith as a religious totem rather than as a communication device—a careful description of the object's attributes should enable another researcher to revise the identification.

## 2. What does it do?

What is its function or use? If it is a mechanism, you can count on the fact that it was probably made for a reason. For what reason was the mechanism made?

If you are not dealing with a mechanism but with an organism, you can identify its behavior. Even a plant like a redwood tree, which takes hundreds of years to grow, moves toward the sun. This positive phototropism can be an answer to the question "What does it do?" Similarly, the behavior of certain compounds, mixtures or elements can be identified under particular conditions. How does this compound behave at STP? How does the behavior change if the pressure is doubled? If temperature is halved?

For all descriptions, be precise and objective in your description. Use standard units of measurement for weight, length, molarity, depth, volume, speed, and so on.

## 3. What does it look like?

Use specific language. Give a picture of the whole first and move to parts. After you have specified the orientation, use the appropriate language. *Dorsal, ventral, anterior, posterior, north, south, perpendicular to, parallel to, intersecting, tangent to* are orientation words.

Once you assign an orientation to the object, your description must unfold with respect to that orientation. Use concrete language.

- **Scale**
When you create a picture in words, give the reader a sense of *scale*. How big is it? How small? Give dimensions in standard units of measurement. A pipet is a suction device, but it can be so small as to measure a pore or so large as to measure a gallon.

Note: Depending upon your purpose and the audience, you may need to devise ways of explaining scale to your reader. For instance, you could tell a lay reader that the Blue Whale grows as long as 98 feet, but how large is a creature 98 feet long? Larger than a dinosaur? Smaller than a football field? Could you imagine a line 98 feet long? But if you were told that you could park a Volkswagen in the heart of a blue whale, that its pulmonary artery was the diameter of the largest garbage can on the market, that a baby whale gains 200 pounds a day for the first six months of its life, you would have a clearer picture of the scale of the animal.

- **Whole to Parts**
If the object has *parts*, describe the parts—after you have given a description of the whole. Pick a point of orientation, and then order the

description of parts with respect to that orientation. Your description *must* be systematic and proceed logically from the orientation. The orientation determines how the subsequent organization unfolds.

Consider the problem of cartographers. How does one render a description of the earth? If the center of the earth is identified as the starting point, the logical sequence of parts would extend from inside to outside, beginning at the center and moving toward successive spherical layers, extending to the atmosphere.

However, if the South Pole were taken as the point of reference, then the subsequent arrangement of continents and seas over the surface would progress with reference to the South Pole. In the middle ages, maps of the world were drawn with Jerusalem at the center and with east at the top, a marked contrast to the contemporary rendering of maps with north at the top. Physicians who attempt to make a diagnosis may classify symptoms from outside to inside, beginning with an evaluation of physical symptoms—demeanor of the patient, color, obvious bleeding, breaks, or dislocations—and leading to an examination of the internal problems as they are manifest through symptoms like fever, leading to a deeper probe with the taking of blood.

## 4. What is it made of?

Identify the material substance the best you can. While you may not know the difference between copper and brass at first glance, you can tell if the substance is made of metal rather than wood, bone, glass, or shell. If the substance is absolutely unknown to you, describe the properties of its substance. At what temperature does this substance melt, freeze, burn. Is it combustible, soluble in water, benzene? Does diamond cut it?

## 5. How does it work?

If the way that an object works is central to the description, organize your description in terms of the operating sequence. What do you do first, then second, then third? In a description of a bicycle emphasizing its operational sequence, start first with the pedal, where pressure is applied; then describe the gears and chain, which distribute the energy received from the pedal; then describe the wheels, which turn in response to the gears; and then the brakes, which stop the wheels.

## 6. How has it been put together?

Unlike the previous category, which deals with an operating sequence, this sequence deals with the way something has been put together. If you have ever had to put a bicycle together for a younger brother or sister, you may have encountered instructions for assembly. The order of assembly differs from the order of operation. Assembly tells how you put something together; operation tells how you use it AFTER it has been assembled. A description emphasizing the order of assembly in a bicycle would start first with the frame, to which is attached (1) the handlebars, (2) seat, (3) wheels. Next comes a description of the brakes,which attach at the handlebars and at points on the wheel.

**Example:** *Spatial (1) and Functional (2) Sequence Combined*

The following passage offers a description of a pipet bulb. This description is called a composite (a whole made of different parts) because it emerged from a collection of student-generated writing. A class was divided into groups and each group was assigned to write a part of the description. Then separate parts of different descriptions were integrated and synthesized to produce the following.

Notice that the title specifies the contents of the report. Although the pipet bulb is used with another tool, the pipet, the pipet is not included in this description.

The purpose and the *audience* were defined for this assignment. The assignment asked the students to imagine that they had invented this object and were submitting this description as part of the petition for a patent. The audience for this paper was the U.S. Patent Office. The purpose was to provide a sufficiently descriptive explanation such that the audience would be able to decide whether this was a new and usable tool.

Notice that the sample gives both a description of the object and a description of the parts based on the way that it would be *used*. The parts were not organized in the way that they would be assembled because this concern is subsidiary to the purpose. If this description were extended, though, to include operation of the pipet with the pipet bulb, a description of the assembly of these parts would be appropriate.

Physical Description of a Pipet Bulb
Introduction

[**Definition**] The pipet bulb is a hand-held laboratory pump (used in conjunction with a pipet) which will extract or release liquids when pressure is applied to the bulb and certain pressure points. [**List of Parts**] The bulb is composed of several parts: a large and small sphere, internal valves which correspond to external pressure points, and stems which connect the parts. [**Operating principle**] As a pump, it operates on the principle of a vacuum.

[**Purpose of the object**] The pipet bulb is used in the laboratory to regulate liquid in a pipet with great accuracy. [**Purpose of the report**] The ease and efficiency of handling make this device unique among other types of hand-pumps, and for this reason we are petitioning for a patent.

## General Description
[**Whole to part**] This one-piece, rubber tool is 20 cm long. Spatially, it may be described in two sections: the large bulb and attachments form the upper segment; the L-shaped branch and small bulb form the lower segment. See Figure 1.

 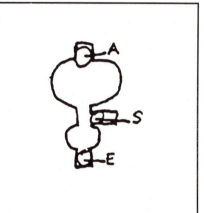

Figure 1: Frontal View of a Pipet Bulb.     Figure 2: Longitudinal Section of Pipet Bulb.

## Description of Parts
[**Organization is from top to bottom and from large to small**] The large sphere is 2.6 inches in diameter and acts as an air reservoir which may be emptied or filled to control the vacuum effect.

[**Description of valves based on use for intake/exhaust**] Air and liquid flow are regulated by the valves, labeled "A," "E," and "S." See Figure 2.

[**Operation**] The two-way valve, "A" (air flow), located at the upper stem which protrudes from the large sphere, regulates air intake and exhaust. Valve "S" (suction), in the lower stem 2 cm below the large sphere, regulates the uptake and expulsion of fluids from the pipette by opening or closing the passage to the vacuum in the sphere. Valve "E" (empty), located 2 cm from the branch of the elbow joint, controls the channel to the outside air.

## Exercises

### 1. Analysis of Two Descriptions: Ostrich and Bird

The following assignment will ask you to compare two descriptions from a sixteenth century book, *On Monsters and Marvels*. *Writing Science* has asserted that good observations, rendered through proper description, retain their merit over time. You will be asked to test this assertion. Using this assertion—adequate writing retains its technically descriptive merit over time—as a hypothesis, test each instance against it in order to arrive at a definition of "technically descriptive merit."

Ambroise Paré (1510–1590) was one of the leading French surgeons of his day. He became chief surgeon to King Charles IX, and then to Charles's successor, Henri III. Paré is credited with two discoveries: (1) pouring boiling oil on a gunshot wound damaged rather than helped the wound, and (2) ligation (tying up veins and arteries) was a better way to stop blood flow than cauterization (applying hot irons) in the treatment of amputations. In his time, Paré was a somewhat controversial figure. Contemporary physicians, who wrote in Latin (as did all scholars, scientists, and learned men of the day), complained that Paré gave away medical secrets because he wrote in French (the common tongue), used pictures, and gave the common name for body parts. "Even women and children can understand his writing," complained the Dean of the faculty of medicine in Paris. (Paré replied that the great authority, Aristotle, was understood in his time by women and girls.)

Paré was also famous for a book, *On Monsters and Marvels*, which represents a then-popular interest in natural history. His book on monsters emerged from his interest in human reproduction: his book begins with an analysis of birth defects. Paré uses *monsters* to describe birth defects, exotic animals, strange happenings, or oddities, curiosities, and hoaxes. Even though his subject matter might seem strange to us, we can appreciate his use of a systematic pattern of order. For the moment, put aside the minor details that might make your reading of these passages strange. Remember that this is a modern English translation of a sixteenth century French book for a popular audience. *Figure* is a term which means form; *portrait* should be understood as *picture* or *diagram*. A span was nine inches; a *doigt* equaled one-sixteenth of a foot. Words in brackets come from the translator who inserted them for clarification. After you read these extracts, answer the questions at the end.

### Flying Monsters
#### Monster 1

This bird is called an Ostrich and it is the largest of all, almost partaking of the nature of four-footed animals, very common in Africa and Ethiopia; it does not budge from land and take to the air, and nevertheless it surpasses a horse for speed. It is a miracle of nature that this animal digests all things indifferently. Its eggs are of a miraculous size, to the point that one can make vases of them; its feathers are very beautiful, as one can recognize and see by this picture.

I do not want to fail to speak, either, of the rarity I saw, concerning the bones of the Ostrich. The late King Charles, having three of them kept at the home of monsieur le Mareschal de Rets, one of which having died, it was given to me, and I made a skeleton [i.e., skeletal diagram] of it. The portrait of which I wanted to insert here, [along] with its description.

A. The head is a little bit bigger than that of a crane, one span from the summit of the head extending to the beak, [the head] being flat, having a beak slit up to about the middle of the eye, this being a trifle round at its extremity.

B. Its neck is three feet long, composed of seventeen vertebrae, which have at each side a transverse apophysis [or process] extending downward, a good inch long, except that the first and second near the head do not have any, and are joined together by ginglymus.

C. Its back, one foot in length, is composed of seven vertebrae.

D. The Sacrum bone is two feet long, or thereabouts, on top of which there is a transverse apophysis beneath which there is a large hole, E, then three other smaller ones, F, G, H; following which there is a box into which the hip bone is insinuated, I, pro-

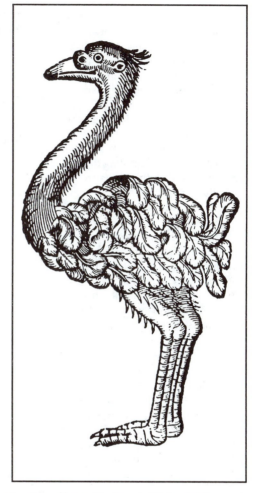

ducing on its lateral external part a pierced bone, K, almost at its beginning, then [it] is united; afterward, said bone forks in two [or, bifurcates], one of which is thicker, L, and the other is lesser, M, each one [being] a half-foot four fingers long: then they reunite, having—between the place where they fork and where they reunite—a hole four fingers wide, N, and longer than a span; then what bone remains is in the shape of a bush hook, or a crooked knife, wide by three fingers breadth, and six inches long, O; then at its extremity it is joined by synchondrosis.

P. The tail bone has nine vertebrae, similar to those of a man.

There are two bones in the thigh, the first of which, Q, the thighbone, is a good foot in length and thick as that of a horse, or thicker; R, the other, which follows it, is a foot and a half in length, having on top a small shank of the length of the bone, losing its pointedness near the bottom.

S. The leg to which the foot is attached is a foot and a half long, having at its extremity two claws [nails, ungues], one large and the other small; there are three bones to each claw [unguis].

V. The bone of the Sternum is of a piece, a foot larger, representing a shield [or, buckler] to which is joined a bone that rides [on] the first three ribs, which takes the place of clavicles.

X. The first bone of the wing is a foot and a half long.

Y. Above it there are two other bones resembling the Radius and Cubitus, at the end of which are attached six bones, Z, which are at the extremity of the wing.

The entire animal is seven feet in length and seven feet and more in height, starting at the beak and ending at the feet.

There are several other remarkable things that I am setting aside for brevity.

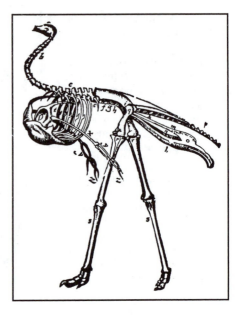

## Monster 2

### From *Monsters and Marvels*

Jerome Cardan, in his book *De Subtilitate* [*rerum*], says that in the Moluccas [or, Spice Islands], one can find on land or on sea, a dead bird called Manucodiata, which in the Indic language signifies "bird of God," which one never sees alive. It lives high up in the air, its beak and feathers similar to the swallow, but adorned with diverse feathers: those which are top of the head are similar to pure gold, and those at the throat to those of a duck; its tail and wings similar to those of a peahen. It has no foot, and if lassitude

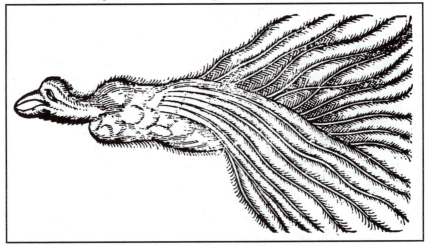

overtakes it, or else it should wish to sleep, it suspends itself by its feathers, which it twists around the branch of some tree. This [bird] flies at miraculous speed and is nourished only by air and dew. The male has a cavity on its back, in which the female broods its young. The interior of this bird, as Melchior Guillaudin Beruce describes, is stuffed and replete with fat, and he says he has seen two of them. As for me, I have seen one of them in this city, that a noteworthy man had, [and] that he held in high esteem: the picture of which you have here.

I have seen one of them, in the city, which was given to the late Charles IX; and I also keep one of them in my office, which [bird] was given to me on account of its pre-eminence.

a. How does Paré organize each description? Look very carefully at the pattern of his ostrich description; Paré's narrative description goes from A to Z; how does this correlate with points on the ostrich skeleton? Compare and contrast the organizational patterns in the descriptions of the two flying monsters.

b. Analyze Paré's use of illustration. Is one more informative than another? Why?

c. Paré claims to have seen both the ostrich skeleton and the "bird of God." Let us suppose that he is an accurate observer: he was, after all, a highly respected attendant to the King. Why does one of his descriptions go against your common sense?

Note: The Manucodiata has been identified as a bird of paradise. Both the bird of paradise and the ostrich are real creatures. This exercise should illuminate an important point. Our understanding of these birds come *through* the description.

### 2. Writing a Description

a. Using the technical description of a pipet bulb as a model, write a technical description of a paper clip, a bow (as in bow and arrow), a straw, a screw, a bobby pin, or a pencil.

b. Let us assume that you have just encountered a variety of a known species or discovered a hitherto-unknown star, element, or compound. Because you claim first discovery, you want to name it. Your name for the thing should observe the rules for nomenclature proper to the field. You must also write an objective description to accompany your claim to naming it. You might explain the reasons for your name in the introduction. Choose from among the following: sea urchin test, Mexican sand dollar, orchid, virus.

c. Write a complete description of a bicycle with the focus on the order of assembly. Then modify your description to focus on the order of operation.

d. Write a description of a shell which will enable your classmates to pick it out of a group of shells.

### References

Lannon, J. *Technical Writing*. Little, Brown, and Company, 1979.
Paré, Ambroise. *On Monsters and Marvels*. Translated and introduced by Janis L. Pallister. University of Chicago Press, 1982.

# 9

❧ ∞ ❧

# Definition

Upon hearing that philosophers in Plato's Academy had defined man as "a featherless biped," Diogenes, the cynic, presented them with a plucked chicken. "Behold! Plato's man!" he said.

Like a Rabbit

On a trip to visit relatives in France, I had the occasion to practice my bad French. My relatives spoke no English apart from "Le Weekend" or "hot dog." One night after we had dinner, I asked my grandmother what was in the *ragoût* (stew). She said it was a *lièvre*. I asked her, "What is a *lièvre*?"

She looked at my uncle. My uncle looked at her. He said, "A *lièvre*—I shot it for dinner." "Yes," I asked, "But what IS a *lièvre*?" My uncle scratched his head, and then he said, "A *lièvre* is like a rabbit but not a rabbit." "How is it like a rabbit?" I asked. He said that it had feet like a rabbit, and ears like a rabbit, but it wasn't a rabbit. Sensing my growing confusion, my grandmother found a French-French dictionary; the dictionary citation said, "*Lièvre*: like a rabbit but not a rabbit." In the English dictionary, a *lièvre* is "Any of various mammals of the family Leporidae, and esp. of the genus *Lepus*, related to and resembling the rabbits but characteristically having longer ears, large hind feet, and long legs adapted for jumping"—like a rabbit but not a rabbit.

A *lièvre* is a hare.

## Words as Concepts: Description versus Definition

A description paints a picture of an object. A definition says what the object *is*. While description and definition seem superficially alike—each renders or represents something—they are conceptually different. A definition says

146

that you *know* what something is: A word can be classified, put in a set, situated within a context. The term locates the concept within a body of knowledge.

Description, in contrast, offers a picture or a representation of something you do not necessarily know. Description creates a picture—or a representation or a model made of words. While descriptions may proceed on the basis of some general knowledge, a term *asserts* a definite identity. While descriptions are built of words, a term is a specialized word which specifies or stipulates the name of a given object, event, or phenomenon.

Chapter 1 made the distinction between (1) what you see and (2) what you think you see. That distinction applies to the difference between description and definition. Description renders *what you see*. Definition identifies *what you think you see*.

In ordinary language, we tend not to pay attention to the strict meanings of words because, in conversation, gestures or facial expressions can supplement language to carry the point, or meaning can be clarified by spoken interchange: "Did you mean this?" "No, I meant that." But in scientific writing, only the words stand before the reader's eyes. Language must therefore be demarcated by the strictest borders of meaning. In the presence of a new concept, old words may have to be redefined. Strictly speaking, to give an object a name is to assert some kind of understanding about the nature of that thing. Further, a term does more than identify a thing, phenomena or event; a term also indicates what the element is NOT. To specify that a creature is a hare is not simply to identify the animal positively, as such; this identification simultaneously conveys the information that the hare is not a rabbit, not a guinea pig, not identical with other things it resembles. In the anecdote above, "Like but not" created a circle of meaning with no center; while the reference to specific parts of the animal might seem to have offered sufficient information, the descriptive detail differed from the specification of an identity. The term *hare* centered the circle of meaning.

## Limiting Alternative Meanings

Scientific terminology should be defined so thoroughly that no other word could match the definition. The definition must fit only the term being defined; a definition must be limiting. *Limiting* means that no other term will satisfy that definition; that definition limits by exclusion all other options. The obverse is also true. No other word can satisfy the definition.

The definition fits only the word; the word fits only the definition; this property of definition is called *reversibility*. Consider the Platonic Academy's definition of man (above).

Man is (=) the only featherless biped.

The term defined is *man*. The definition is "featherless biped." Could *man* be anything other than a featherless biped? Could a featherless biped be anything other than a man? Diogenes, cited in the opening to this chapter, challenged the reversibility of the definition. By producing a chicken, he showed that the definition was not sufficiently limiting. Man is not the only featherless biped. A plucked chicken also matches that definition.

- Write to exclude alternative meanings.

## Kinds of Definitions

A definition may make one of two assertions; a definition may assert: (1) "This is how everyone uses the word" or (2) "This is how I am now using the term." The historic-authoritative definition[1] (1) emerges from a common language of science, through its use by authorities in the field to indicate what the generally conceived meaning is. The stipulative definition (2) offers a meaning that an individual proposes. A stipulative definition is specific to context; it does not have a universal meaning commonly recognized by all who share the scientific body of knowledge. When one coins a new term or provides a new meaning to an old term, one provides a stipulative definition. The historico-authoritative definition specifies: "This is *what is meant* by the word." The stipulative definition asserts "This is *what I mean* by the word."

As you read scientific papers or other kinds of writing, be alert to the nuances in the status of meaning: whether the definition is being reported as factually true or whether the writer offers a provisional or assigned meaning. Be aware of this distinction when you write definitions as well. Note, however, that the distinction between a historically received definition and a stipulated definition may not be so clear cut as represented here.

The following points apply to all kinds of definition. However they may vary in form or use: (1) because you want to make yourself understood, your definition should be less complicated than your term; (2) your definition should in some way define the thing by noting its attributes (if a substance, its color, matter, form, material, etc., or, if a process, the conditions under which the process takes place); (3) you should always write in plain English; (4) a definition should not use figurative language; (5) the definition should be stated in positive terms.

While there are many ways to define words, the rest of this chapter will explain the function and method of writing three kinds of definition. The *form* of these definitions can be used for both stipulative and historico-authoritative situations of use. They are (1) *parenthetical* definition, (2) *sentence* definition (Aristotelian definition), and (3) *extended* definition.

Each type of definition has a different form and a different use. The *parenthetical* definition is a word or phrase, expressed at the same level of generality as the term, which is integrated with the sentence. It is short, frequently synonymous, and does not disrupt the flow of the writing. A *sentence* definition gives a longer explanation expressed in a complete sentence or explanatory phrase. It is found in the body of your report or in a glossary. It may be the basis of further discussion or the foundation for an extended definition. If your paper gives greater emphasis to a word, or if a word needs to receive more explanation because the idea is important and you want to show that importance, the *extended* definition, which may range from a paragraph to one or more pages may be the best choice.

## Parenthetical Definition

The term *parenthetical definition* is precisely descriptive; you place a parenthetical definition within the parenthesis in the body of your text. *Parenthetical* is the adjective form of *parenthesis*. This type of definition refers to a synonym in parenthesis immediately following the term it defines; it can also be a phrase or explanation stated at the same level of generality. It may sometimes be the word in Latin, the genus species name, or units of measurement. The parenthetical definition is integrated with the text. For punctuation purposes, you should consider it as part of the term it explicates and punctuate after the parenthesis.

Sample Parenthetical Definitions
1. Unlike the thread of the silk moth larvae (*Bombyx mori*), spider silk has not yet been harnessed for mass production.
2. We are often aware of the rhythm of the heart. It works hard, pumping about 70 barrels of blood a day, alternating rhythmically between dilation (diastole) and contraction (systole).
3. One volunteer lived in the luxurious underground apartment for eight days in dim illumination (40 lux) and showed an activity-sleep cycle with a 24.5-hour period.
4. People lacking vision from early youth have shown disorder in water balance (nycturia) and glucose balance.
5. The spacecraft belonged to the National Aeronautics and Space Administration (NASA); NASA also sent up the Jupiter probe.
6. The effervescent (bubbling) mixture was quickly discarded.
7. The leaching field (sievelike drainage area) needs fifteen inches of crushed stone.

## Sentence Definition

Sentence definitions are more expanded than synonyms. Sentence definitions are used (1) in the introduction of your paper, (2) and/or in a glossary which may accompany your paper, and (3) they may be also form the foundation for an extended definition. Unlike a parenthetical definition which does not disturb the flow of your writing, a sentence definition gives the term greater emphasis (because it is a whole sentence rather than a synonym). If you wanted to assign even greater priority to the term—if your entire paper or a segment of your paper were devoted to it—you could use an extended definition.

The word to be defined is called the *term*. The group of like items to which it belongs is called the *class* (category). The name of the class derives from the common traits belonging to the members of the class. The *differentiae* tell how the term differs from other items in the class. To write a sentence definition, specify the term, classify it (the class tells what the items have in common), and then specify the differences (tell how the item is different).

## Formula for Sentence Definition

| Term | = (is) | Class | + | Distinguishing features (differentiae) |
|------|--------|-------|---|----------------------------------------|
| tepee | = | a form of portable dwelling | + | cone-shaped (with a hole in the top); made of skins and sticks; used by Plains Indians; transported by pack animals |
| vibrato | = | a musical embellish-ment | + | that depends primarily on periodic variations of frequency which are often accompanied by variations in amplitude and waveform |
| charm | = | a quantum number | + | which has been proposed to account for an apparent lack of symmetry in the behavior of hadrons relative to that of leptons, to explain why certain reactions of elementary particles do not occur, and to account for the longevity of the J particle. |

## To Make a Sentence Definition . . .

1. Put the term in a group with other items or concepts like it. In other words, classify your term.
2. Name the class.

3. Explain, in positive terms, how your term differs from every other item in the class.

## Discussion

1. All sentence definitions must include classification. No single classification is intrinsically "right"; a classification may be appropriate or inappropriate, the definition of propriety hinging upon the writer's purpose and audience. Further discussion of classification follows (below).

   The *audience* and context define how general or how particular your classification should be. You can decide how to classify a term on the basis of your purpose.

2. Your term must be completely distinguished from every other item in the class. In other words, your definition *excludes* alternative meanings. It must apply to that term and that term only. It must satisfy the principle of reversibility. If another term could be defined by that definition, your definition is incomplete.

3. Be objective. Use a category your audience will recognize. If you rely upon a personal system of categories which derive from your value system, the communication scene will be compromised. Avoid classifying a term by highly subjective criteria, as in *favorite* or *important*. For example, "Free enterprise is the best economic system" expresses an opinion.

4. The differentia must sufficiently distinguish the term from every other item in the class. If the selected category is very general, the distinguishing detail will be extensive. If a horse is classified as belonging to the category of animate beings, the detail carry the burden of specifying how the horse differs from a zebra, from a bird, from a fish, from an amoeba, from a flower, from a fungus. In this case, a more limiting category—odd-toed herbivores for instance— will also limit the amount of detail needed to distinguish the horse from other beings.

5. Explain the difference in positive terms. When you explain how the term differs from other elements in the class, do not express this in negative terms. Consider the definition of tepee (above). To state that a tepee is a form of portable dwelling which is not a tent fails to provide adequate differentiation. State the positive attributes of the terms which distinguish it from others in the class: The tepee is

composed of animal hide and sticks (versus the tepee is not made of metal).

## The Significance of Classification

Many words possess more than one meaning; the specific meaning is governed by context. For example, *mouse* can refer to the computer mouse which belongs to the set of manipulating devices [mouse, joy stick, keyboard], or *mouse* can refer to a biological organism included in the set of rodents [mouse, rat, guinea pig].

Classification answers the question: *what kind of?* Classification imposes a context by specifying the category of things. Context (classification) rules out the alternative meanings of the word; thus, specifying the context (class) enables you to limit alternative meanings. The *McGraw-Hill Dictionary for Scientific and Technical Terms* (2d edition) provides the following series of definitions for *shift*.

[Automatic data processing] A movement of data to the right or left, in a digital-computer location, usually with the loss of characters shifted beyond a boundary.

[Geology] The relative displacement of the units affected by a fault but outside the fault zone itself.

[Industrial engineering] The number of hours or the part of the day worked. Also known as tour.

[Mechanical Engineering] To change the ration of the driving to the driven gears to obtain the desired rotational speed or to avoid overloading and stalling an engine or motor.

[Metallurgy] A casting defect caused by malalignment of the mold parts.

While the term (*shift*) remains the same for all these definitions, the various contexts—data processing, geology, industrial engineering, mechanical engineering, metallurgy—make the word mean different things.

## Possible Pitfalls: Do Not Fall into the Fallacy

### Genetic Fallacy: Eggs to Chickens

A genetic fallacy is an error in reasoning (fallacy) in which one defines a term by its origin or genesis. This fallacy is called *genetic* because *genetic* is the adjective form of *genesis*. Only in special cases (medical textbooks, for instance) is it appropriate to specify a medical pathology by reference to the organism that causes it. In most cases, it is insufficient to identify a term *solely* by its origin. (This problem is a type of incomplete differentia.) For example, the statement

- A chicken is a creature which hatches from an egg

offers an insufficiently limiting definition. Among its problems, the generality of the class requires more differentiae to distinguish the chicken from other creatures similarly hatched. The origin is not sufficiently differentiating because other creatures hatch from eggs: reptiles, amphibians, other birds. Even if the class were modified—"A chicken is a flightless bird which hatches from an egg"—the problem of insufficiently limiting differentia remains, as the definition does not exclude all alternative meanings.

### Tautology or Circular Definition

In your sentence definition, the term should not appear twice in the definition. Never use the term, a version of a term, or a synonym for the term within a sentence definition. *Tautos* means the "same": A tautology or tautological reasoning means (1) the term appears on both sides of the equals sign, or, more broadly, (2) you assume or take for granted the idea you want to define. For example, "The Carnot cycle is a cycle named for Carnot" defines the term in terms of itself. It is circular.

The best definition frames the term from within a context or class more general than the term; the specifying detail is phrased more particularly than the term itself.

Whole arguments are vulnerable to circular reasoning. When this fallacy is extended throughout an argument, the point is presented as evidence of itself, or the point which requires support is used to support itself.

### Phrasing

The definition must be grammatically correct and grammatically parallel. If the term is phrased as a noun, the class must be phrased as a noun. For example, the definition—"Skiing is when you ski down a hill"—is not only circular, it is also grammatically incorrect. "When you ski down hill" is a subordinate clause which, by the rules of grammar, cannot serve as the predicate noun. The classification should read "skiing is a sport" not "skiing is when . . ."

## Summary of Criteria for Good Sentence Definitions

1. A good sentence definition contains all three elements: term, class, and differentiae.
2. The definition must be logically consistent, employing several levels of generality; it must not be circular.
3. The definition must be objective, not subjective.
4. The definition must be suited to audience and purpose.

5. A sentence definition is exclusive. It defines the specified term and no other.

## Exercises in Sentence Definitions

### 1. Incomplete definitions
Many of the following are not complete definitions. (1) *Identify* the problem in each of the following and then (2) *rewrite* the definitions to complete them.

1. A fish-finder finds fish.
2. A point is that which has no part.
3. The icosahedron is a Platonic solid.
4. The wings form an important part of an airplane.
5. A pandemic is like an epidemic.
6. Rain is made by liquid water drops with diameters greater than 0.5 millimeter; or if widely scattered, the drops may be smaller.

### 2. Write your own definition
a. Write a sentence definition for one of the following: *desert, pipet, torque, contract, tension, concrete, byte, hypertext, screw.*

b. Without a dictionary, write as many sentence definitions for the word *file* as you can think of. Then list one synonym for each of those terms. Try the same exercise with *stress, work, field.*

### 3. Definition for Glossary
Read the following passage. As part of the glossary which will accompany your formal report, write a sentence definition for *whole mouse fluorography.*

A variety of established methods was employed in exploring this issue, and a new method for the analysis of the patterns of DNA synthesis in sectioned whole mice was developed, termed whole mouse fluorography. This newly developed protocol involving thin sections of whole mice, previously treated with triated thymidine, was used to establish the patterns of cellular DNA synthesis in the organ system of a newborn mouse. Our results indicate that acute polyoma virus infections appear to have no substantial effect on these patterns. The absences of tritium signal from bodily fluids (blood, urine, etc.) further indicates that the images present by whole mouse fluorography do correspond to patterns of cellular DNA synthesis resulting from the incorporation of 3H-thymidine. —Juan Moreno, *Journal of Undergraduate Research in the Biological Sciences* 21 (1991).

# Extended Definition

In the body of a report, you may find that you need to give your audience more information than simply a sentence or parenthetical definition. You may, in fact, want to focus your entire report on offering a definition, either of a new technology or procedure, or of an organism.

How to extend your definition THROUGH . . .

## 1. Negative Definition

*Negative definition* names what a thing is not. It makes explicit what is normally taken for granted.

Topology (a branch of mathematics) studies the properties that remain unchanged when shapes are deformed by twisting or stretching or
squeezing. Whether a shape is square or round, large or small, is *irrelevant*
in topology because stretching can change those properties. Topologists
ask whether a shape is connected, whether it has holes, whether it is
knotted. They imagine surfaces *not* just in the one-, two-, three-dimensional universes of Euclid but in spaces of many dimensions, impossible
to visualize.[2]

## 2. Etymology

*Etymology* means the origin of the word (the history of the *word* is different
from the history of the *thing*.) Frequently the history of the word (its origin,
change, development) will give you information about the term to which
it applies. When you explicate the letters in an acronym (a word made from
the first letter of other words), you will learn more about the thing. SCUBA
comes from Self Contained Underwater Breathing Apparatus. NASA refers
to the National Aeronautic and Space Administration. The word *vaccination*
comes from *vacca*, the Spanish word for cow; the history of this word will
open up the history of the process since it began with Jenner and his cowpox
vaccine. *Cytokinesis* means "movement of the cytoplasm." Humoral immunity results in the production of antibodies which circulate in the blood-
stream; *humor* comes from the Latin word for fluid.

## 3. History

A history or the background of the item provides your reader with additional information. If writing about a mechanism or procedure, you might
supply information about its discovery, development, method of manufacture, changing applications, and possibilities for use in other fields. A
discussion of genetics might lead you to discover an abbey garden, where
in the 1860s, an Augustinian monk, Mendel, discovered the principles of
inheritance. The following paragraph describes the history of influenza
epidemics.

The highly contagious acute respiratory illness known as influenza
appears to have afflicted humans since ancient times. The sudden appear-

ance of epidemics of respiratory diseases that persist for a few weeks and equally suddenly disappear is sufficiently characteristic to permit identification of a number of major epidemics in the distant past. One such epidemic was recorded by Hippocrates in 412 B.C., and numerous episodes were described in the Middle Ages. Webster (646) and Hirsch (266) collected historical data from 1500 until 1800.[3]

## 4. Example

An example offers a way to extend your definition through particular instances. The example offers a concrete embodiment of the principle or general point. Examples should be geared to the interests of your audience.

- *Shamu* and *Orca* are both particular instances of killer whales.
- The order Perissodactyla are herbivores; they possess hooves with an odd-number of toes on each foot; the *horse* and *zebra* are examples of the order Perissodactyla.

## 5. Graphic Illustration

A graphic could be a photograph or a diagram which shows the object from a specified point of view: cross-section, lateral, exploded. A graphic organizes information visually. Some graphics encode a dense amount of information, for instance, a sine wave chart; other graphics handle less dense quantities of information: A pie chart spatializes the division of a whole into parts in order to illustrate relative size. Your graphic should be labeled; if it is borrowed, credit your source at the bottom left corner of the frame. Make sure that you integrate the graphic with the text; the graphic does not stand by itself but is included within the body of your definition. Refer to the graphic in your writing. If you need to explain your graphic, put the explanation after it. See Chapter 7, Graphics, for further elaboration. An example of a graphic[4] follows.

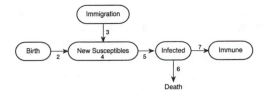

## 6. Description/Analysis of Parts

Static objects, moving objects, concepts or processes are comprised of component parts. An effective way to extend a definition is to break the object into parts and then (1) describe the parts and (2) explain how the parts fit together. If you deal with a static object, use a consistent pattern of spatial description: from whole to part; part to whole; top to bottom, inside to outside, most important to least important. If explaining a process, be consistent about organizing the temporal pattern: from first to last, last to first, most important to least important. See Chapter 8, Description, for further elaboration.

> An antibody molecule is made up of four proteins that are bound together: Two of these proteins are identical and are called *heavy chains*; the other two are also identical and are called *light chains*. A protein is a linear chain of building-block molecules called *amino acids*—much like beads on a string. There are 20 possible amino acids, and the nature of a protein is determined by the particular sequence of the amino acids it contains. In the case of antibodies, the two heavy chain proteins are larger than the two light chain proteins. These proteins are held together by chemical bonding in a Y-shaped molecule. . . . Each antibody molecule is specific for one particular antigen, and this specificity is determined by the sequence of amino acids in the heavy and light chains.[5]

### 7. Comparison/Contrast

Comparison refers to the resemblance or similarity between things; contrast refers to differences. Comparison/contrast extends a definition by clarifying what things your term resembles or differs from. See Chapter 9, Comparison/Contrast, for further discussion.

There is no comparison without contrast; there is no contrast without comparison.

### Likeness

(1) Spider and silkworm are not at all closely related, yet *their silks are similar. Both use keratin*, a protein that is found in hair, horn, and feathers.

### Difference

(1) Yet the silk of the spider's web has many fabulous qualities. *It is much stronger than the silk of a silkworm.* Moreover, *it is variable*: Spiders can produce different kinds of silk for different functions.[6]

### 8. Operating Principle/How It Works

Mechanisms, static or dynamic, work according to a principle whose explanation should be part of your definition. For example, a paper clip is a form of compressed spring; a bow, too, works on the principle of a spring by which energy is stored and released through mechanical means. Even biological or chemical processes can operate by a general guiding mechanical, energetic, or physical principle.

Capillarity is the action by which the surface of a liquid where it contacts a solid is elevated or depressed, because of the relative attraction of molecules of the liquid for each other and for those of the solid. —*McGraw-Hill Dictionary of Scientific and Technical Terms*, 2d. ed.

## 9. Special Conditions or Qualifications

A detailed definition should include information about special circumstances, as they obtain. It qualifies the information by specifying conditions, limitations, or qualifications.

Ultimately it may be possible to use these [blood cell] growth factors to stimulate and regenerate the immune system in AIDS patients. Of course, it will be important to use these growth factors in conjunction with antivirals. Otherwise, continued HIV infection would destroy the immune system again.[7]

# Examples of Extended Definitions

In the following passages, study the use of the different ways to extend a definition.

### Example: Arch Dams

[Sentence Definition] An arch dam is a curved dam that carries a major part of its waterload horizontally to the abutments by arch action. [Special case: exception] Massive masonry dams, slightly curved, are usually considered as gravity dams, although some parts of the loads may be carried by the arch action. [History] Many early dams were built of rubble, ashlar, or cyclopean masonry. [Contrast] However, practically all arch dams constructed during recent years have been built of concrete.

[History] Arch principles have been used in bridges and buildings since about 2000 B.C. [Examples] Apparently Ponalt dam, built in Austria in 1611 A.D., was the first arch dam recorded in engineering history. The 64-ft Bear Valley Dam, built in the San Bernandino Mountains in southern California in 1883, was the first arch dam constructed in America. It was followed by the 95-ft Sweetwater Dam, in 1888, and the 88-ft Upper Otay Dam, in 1900, both built near San Diego, California. Since 1904, many arch dams have been built in the United States and abroad. —Ivan E. Houk and Roman P. Wengler, *Handbook of Applied Hydraulics*, 3d ed., ed. by Victor Calvin Davis (New York: McGraw Hill, 1969), section 14.

Example: Variable
Student-Authored Extended Definition
Definition of "Variable"

Audience: Lay audience

Purpose: To give an extended definition of the word *variable*

[Sentence definition] A variable is a value, constant, condition, or set of conditions that are subject to change and are often assigned to a particular letter or word. A variable is characterized by the variable itself and its equivalent of the constant, condition, or set of conditions that it represents. [Etymology] The word "variable" derives from its root word "vary" which means change or changing. [Purpose] The main use of a variable is to simplify the process of manipulating and monitoring a constantly changing value or condition(s). [Negative definition/comparison/contrast] A variable, by definition, is not a constant (remaining the same without change); it is a continuously updated (changed) according to the alterations made by the system in which the variable is used. [Analogy] A variable, in an abstract analogy, is similar to the wind. As the wind constantly changes in both velocity and direction, a variable constantly changes to represent an updated or changed value or condition(s). [Division into parts] There are several varieties and types of variables. One type is a string (a sequence of letters such as a password); the other is a single letter variable (like the letter v). [Special conditions] Special conditions apply to the word variable when it is being used in the context of computer programming. For example, a computer can be told to interpret (understand) a variable, manipulate a variable, assign a new variable to present the outcome of manipulations (such as calculations). In this case, a hierarchy of variables is formed and a new variety of variables is born. Variables are then the means through which a computer, or any other such interpreting device, can keep itself updated to current conditions, thereby making itself more applicable to present problems and conditions. See example.

[Application] An example of variables used in math:

Let x=time to travel from point A to point B.

Let y=time to travel from point B to point A.

x + y = total time traveled.

—Andy Burchett, CAMP student, UCI, summer, 1992.

## Definition and Introductions

A definition offers a solid contact point between the reader and the writer.

To start a paper with a definition is to take control of specifying the context for your topic. Notice in the following excerpts how the writer starts with a definition in the introduction and then moves from general discussion, set out by the definition, to the particular objective of the paper. The following were taken from student papers prepared for an upper division writing course.

1. Photodynamic Therapy (PDT) is a promising modern experimental technique used for the treatment of various malignancies (1, 2, 5, 17). The process of PDT involves the utilization of a photosensitizing dye which selectively localizes in cancerous tissue (2, 1, 4); the tumor is destroyed when laser light with sufficient energy is tuned to the appropriate wave-length of the photosensitizer (3, 1, 3). When photosensitizer molecules become excited by the energy of the laser light, high reactive singlet oxygen is generated (111). Singlet oxygen plays a major role in the tumor destruction of PDT by interacting with cellular components and tumor vasculature (111). —Cesar Zamora, "Photoxicity Comparison of Phthalocyanines."

2. Zoopharmacognosy is the scientific study that describes the process by which wild animals select the use of plant-derived properties (i.e., leaves, stems, roots, resins, etc.) with medicinal properties for treatment and prevention of disease. Coatis (*Nasua Narica*) of the island of Baro Colorado have been observed utilizing a resin from the tree *Trattinnikia Aspera* as a topical agent. —Arturo Pelayo, "The Chemistry of Resins used by Coatis in Fur Rubbing."

3. Artificial intelligence (AI) is part of computer science that deals with symbolic representations of knowledge. AI tries to build intelligent machines which interact with the environment. It is used mainly in robotics, natural language understanding, speech understanding, vision, general problem-solving, and other areas. —Martin Raymundo, "Artificial Intelligence and Its Applications," paper prepared for CAMP, 1995.

4. Superoxide dismutases are the metalloenzymes identified as part of an endogenous antioxidant system present in various eukaryotic organisms. —Linda Tesoriero, from abstract to "Resistance of Adult Drosophila Melanogaster Females with Different Superoxide Dismutase Genotypes to Irradiation."

5. The nucleus of the optic tract (NOT) is part of the pretectal complex and involved with the detection of retinal slip, a gaze resulting in

an involuntary movement of the eye to follow an object detected by the retina, which is the stimulus for generating horizontal optokinetic nystagmus (hOKN) (Simpson, 1988). The hOKN is a set of eye movements caused by motion of the entire visual field and stabilizes the eyes when the subject itself is moving. —Evelyn Tom, "Cortical Projections in guinea Pig, to the NOT, Area responsible for Optokinetic Nystagmus."

## Exercises

### 1. Analysis of Extended Definition

Read the following extended definition. Then (a) locate the sentence definition, (b) identify elements of the extended definition, (c) make a reasonable interpretation about the audience level.

> The propionibacteria are pleomorphic, nonsporeforming, gram-positive rods that frequently form irregular clumps with "Chinese-character" arrangements. They are nonmotile, anaerobic to aerotolerant mesophiles that form small raised colonies that are cream, yellow, orange, or deep red. The overriding characteristic that gives them their name is the production of large amounts of propionic and acetic acids. These acids can be produced by fermentation not only of various sugars, but also of lactic acid. Yields of ATP are low in these pathways, and growth is slow, particularly on solid media. This characteristic of slow growth is of practical significance in the laboratory and industrial applications; visible colonies may take from 5 to 14 days to appear, depending on the composition of the medium. —Bonnie Glatz, "The Classical Propionibacteria: Their Past, Present, and Future as Industrial Organism," *ASM (American Society for Microbiology)* 58, No. 4 (April 1992): 197–201.

### 2. Definition of Contagion

Turn to Girolamo Fracastoro's "The Germ Theory of Disease" in the appendix and analyze Fracastoro's use of extended definition.

### 3. Stipulative Definition: Hippocrates

Turn to Hippocrates' "Epidemics, Book I" in the Reading Selections and write a stipulative definition for *causus*. Further instructions are located at the end of that reading.

## References

Corbett, Edward. *Classical Rhetoric for the Modern Student.* Oxford University Press, 1990.

Fan, H., Conner, R., Villarreal, L. *The Biology of AIDS*, 3d edition. Jones and Bartlett Publishers, 1989, 1991, 1994.

Glatz, Bonnie. "The Classical Propionibacteria: Their Past, Present, and Future as Industrial Organism." *ASM (American Society for Microbiology* 58, No. 4 (April 1992): 197–201.

Hanson, N. R. *Perception and Discovery: An Introduction to Scientific Inquiry*. Freeman, Cooper, and Company, 1969.

Jason, G. *Introduction to Logic*. Jones and Bartlett Publishers, 1994.

Lannon, J. *Technical Writing*. Little, Brown, and Company, 1979.

Daniel Lapedes, ed. *McGraw-Hill Dictionary of Scientific and Technical Terms*, 2d ed. McGraw-Hill, 1974, 1976, 1978.

## Notes

1.  This term is taken from Norwood Russell Hanson's *Perception and Discovery: An Introduction to Scientific Inquiry* (San Francisco: Freeman, Cooper, and Company, 1969). Other philosophers and logicians may propose other terms. For instance, Gary Jason, in *Introduction to Logic* (Boston, London: Jones and Bartlett Publishers, 1994) calls this kind of definition *reportive* (p. 180).

2.  James Gleich, *Chaos* (New York, London: Penguin Books, 1987), p. 46.

3.  Brian R. Murphy and Robert G. Webster, "Influenza Virus," *Virology*, edited by B. N. Fields, et al. (New York: Raven Press, 1985), p. 1179.

4.  Hung Fan, Ross Connor, and Luis Villarreal, *The Biology of AIDS* (Boston, London: Jones and Bartlett Publishers, 1994), p. 10.

5.  Fan, et al., *The Biology of AIDS*, pp. 33–34.

6.  Fritz Vollrath, "Spider Webs and Silks," *Scientific American* (March 1992): 70–76.

7.  Fan, et al., *The Biology of AIDS*, p. 154.

# 10

───── ∾⤫∾ ─────

# Classification as Context

Classification refers to the act of putting elements in a group based on similarities; this is the same operation in the making of sets. For the purpose of this discussion, *set* is another word for *class*.

A fundamental process in cognition and an important ordering method in writing, classification imposes order on animate and inanimate objects, states, and phenomena. Classifications do not occur in nature; human beings identify patterns and impose categories on them. The classification of phenomena leads to the formulation of concepts and definitions. Classifications are important in science because they enable the codification and regularization of discrete data. Classification, by asserting a relationship of likeness between elements included in a class, describes patterns.

- What's wrong with this picture? (Circle the element that doesn't fit in.)

[peso, lira, dollar, pound, tobacco]
[hydrogen, carbon, methane, oxygen]
[fish, whale, dolphin, porpoise]

In the first instance, *tobacco* does not fit into the group of items because three of the four items in the set are types of currency. *Methane*, a molecule composed of carbon and hydrogen, differs from the other items which are single elements. While fish and dolphins are marine animals, *fish* is the only non-mammalian listing in a series otherwise composed of marine mammals. Most people can recognize a pattern and spot the variation in the pattern; science involves working with the most generally understood and recognized types of pattern. Pattern recognition, whether intuited or explicated, forms the foundation for the operation of classification.

## The Importance of Classification

Classification is important because it gives us a perspective on the object in question. Classification defines what we see because it determines HOW we see. HOW we see defines the way we TREAT the object in question. Classified as a source of animal products, whales at one time were killed for dog food; classified as an endangered species, whales are protected from commercial hunters.

Most states require testing for sexually transmitted diseases (STDs) before marriage. All states have laws requiring public health care agents to report STDs to the health department. In the matter of STDs, the individual's right to confidentiality is superseded by rule of law. While sexual intercourse offers one route of the transmission of AIDS, most states have not classified AIDS or HIV+ status as an STD, and hence, the disease does not legally fall under the laws which require reporting.[1] While some states have laws requiring public-health reporting, this decision is made at the state level not the national level. If the disease were to be formally classified as an STD, then the way that cases are treated would be subject to regulation at a national level.

S. I. Hayakawa, a linguist and former Senator from California, had this to say about classification:

> Most intellectual problems are ultimately problems of classification and nomenclature. Some years ago there was a dispute between the American Medical Association and the Antitrust Division of the Department of Justice as to whether the practice of medicine was a "profession" or "trade." The American Medical Association wanted immunity from laws prohibiting "restraint of trade"; therefore, it insisted that medicine is a "profession." The Antitrust Division wanted to stop certain economic practices connected with medicine, and therefore it insisted that medicine is a "trade." Partisans of either side accused the other of perverting the meanings of words and of not being able to understand plain English. . . . Is aspirin a "drug" or not? In some states, it is legally classified as a "drug," and therefore it can be sold only by licensed pharmacists. If people want to be able to buy aspirin in groceries, lunchrooms, and pool halls (as they can in other states), they must have it reclassified as "not a drug."
>
> In any case, society as a whole ultimately gets, on all issues of wide public importance, the classification it wants, even if it has to wait until all the members of the Supreme Court are dead and an entirely new court has been appointed. *In short, society regards as "true" those systems of classification that produce the desired results.*
>
> The scientific test of truth, like the social test, is strictly practical, except for the fact that the "desired results" are more severely limited. The results . . . decided by scientists are only that our systems produce predictable results. Classifications, as already indicated, determine our attitudes and behavior toward the object or event classified. When lightning was classified as "evidence of divine wrath," no courses of action other than prayer were suggested to prevent one's being struck by lightning. As soon, however, as it was classified as "electricity," Benjamin Franklin achieved a measure of control over it by his invention of the lightning rod. Certain

physical disorders were formerly classified as "demonic possession," and this suggested that we "drive the demons out" by whatever spells or incantations we could think of. The results were uncertain. But when those disorders were classified as "bacillus infections," courses of action were suggested that led to more predictable results.

Science seeks only the most generally useful systems of classification; these it regards for the time being, until more useful classifications are invented, as "true."[2]

## Making Interpretations

Classification is a straightforward operation but becomes problematic when creatures or phenomena fall between the borders of established categories. What happens when a subject has traits that belong to more than one class? How do you decide which class it fits into?

California Sheephead are a type of fish which, in the course of their lives, take on both sexes. They are female until they are seven or eight years old and then the ovaries become testes, and the female becomes male. What would be the gender of a sheephead at the age of seven and one-half?

Is a virus more like a living creature or more like a non-living creature? Is a platypus more like a bird or more like a mammal? In determining which criteria are most applicable, you have to make a reasoned interpretation. Very frequently in science, you will find that scientific innovations come about because the classification of data changes or the applicable set of criteria shifts. *The data stays the same but the interpretation of the data changes;* this is has come to be called a "paradigm shift." Since we look at the world through certain models or structures, what Kuhn called a paradigm, a change in perception can bring about a corresponding change in interpretation. Many changes in science came about as a result of a change in paradigms.

Intransigent elements—whether they be the platypus, the virus, or light particles—offer an important path way to changes in science. The behavior of the virus enabled us to raise two questions: (1) Were the current modes of classification sufficient to account for an organism that manifests traits of non-living and living beings? (2) Did this organism require a new taxonomy to accommodate it?

Be alert to the way that classification defines what and how you see. Do not feel that you have to automatically accept orthodox classifications. In your mind, ask questions.

### How to Classify

1. Put like items together. Make a set. Group like with like.

2. Name the group. (Name the set. The name of the set differs from the items in the set.
3. Identify the principle which defines the inclusion or exclusion of other items from the class.

## 1. Group Like with Like.

[Ford, Chrysler, Mazda, Toyota] comprise a set of automobile types. This set could be named a class of automobile manufacturers. You could take any item from this set, Toyota, for instance, subdivide it, and create another set on the basis of kinds of cars: Toyota makes [four-door sedans, two-door sedans, sports coupes]. You could take "sports coupe" and, in turn, make it the name of a set whose elements include [MG, Fiat, Mazda].

Cars are an easy choice here. But how would you classify a virus? In the early history of microbiology, scientists were tentative about specifying the nature of viruses because viruses exhibit the behavior of both living and non-living things. Similarly, the duckbill platypus initially posed a problem for taxonomists because it possessed some traits specific to birds and some traits specific to mammals: like a bird, the platypus has a beak and produces young by laying eggs; like a mammal, the platypus is warm-blooded and suckles its young. To which group did the platypus belong?

The platypus and the virus are not the only creatures that represent elements that straddle the border of conventional categories. The prehistoric creature, *Archaeopteryx*, seems part bird and part lizard; how would it fit into our understanding of evolution? Was light a particle or a wave? Was electricity a type of fluid or a type of air? Classification is important, in each instance, because the class prescribed the kind of physical laws that could be applied to it.

When Newton derived his understanding of gravitation, he recognized a pattern. He classified two previously unconnected motions as expressions of the same pattern: the falling of an apple with the orbit of the moon.

## 2. Name the set.

Naming the set is important because it tells the reader what you think the items have in common. The name of the set is not included in the set; it exists as a broader and more comprehensive element than the set. The name of the set helps the reader to understand what the items have in common. The name of the set is a name of a whole; the elements in the set are the parts of the whole.

## 3. Specify the ordering principle.

The ordering principle specifies what the set items have in common. Set A—[2, 4, 6, 8 . . .]—is the set of even numbers; even numbers is the ordering principle, the principle that establishes the pattern. The principles which describes the common elements in the set can also function as the principle which is invoked to determine if other elements outside the set fit into the set.

For example, does Set B—[3, 5]—fit into Set A? Because Set A= even numbers, and Set B contains odd numbers, the ordering principle prohibits this inclusion. The ordering principle is the rule which you apply to other items in order to determine whether or not they fit into the already-formed set. Take, for instance, the duckbill platypus. This creature has the beak of a bird—thus its name *duckbill*—but it also possesses hair, suckles its young, and is warm-blooded—all the traits of a mammal. Which set of criteria—those which define "bird" or those which define "mammal"—applies to the duckbill platypus?

Note: The elements in a set should have more in common than simply their inclusion in a class. For instance, take the following items:

Set A [ultra-violet, hoof, Borges, pi]

What do the elements of Set A have in common? Is there anything which connects ultra-violet, hoof, Borges, and pi? There is no discernible pattern of similarity apart from the fact that they are all included in Set A. Set A is a class of miscellaneous elements.

### Extended Example

Among the many accomplishments of the Italian painter Leonardo da Vinci (1452–1519) was his detailed anatomy studies. The following passage, an excerpt from his anatomy notes, shows how he organized his prospective work through classification. He planned to make annotated sketches of the following:

(A) Man. The description of man, which includes that of such creatures as are almost of the same species, as apes, monkeys, and
the like, which are many.
(B) The Lion, and its kindred, as panthers, ounces, tigers, leopards,
wolves, lynxes, Spanish cats, common cats, and the like.
(C) The Horse and its kindred, as mule, ass, and the like, with incisor
teeth above and below.
(D) The Bull and its allies with horns and without upper incisors, as
the buffalo, stag, fallow deer, roebuck, sheep, goat, wild goats,
musk deer, chamois, giraffe.[3]

What do the items in each set have in common?
If you study Leonardo's classification carefully, you might notice that he places "wolf" in the same category as "tiger" (Set B), and he places

"giraffe" in the same category as "sheep" (Set D). The reasons for the classification in Set B may not be entirely clear to us. In fact, Set B may look something you've seen on a test:

Which one of these does not fit into the group:

(1) tiger, (2) wolf, (3) lion, (4) common cat

If you had seen this list on a test, you could answer that "wolf" (2) does not fit into this list because all of the other animals are felines and a wolf is not a feline. But you could also answer that "common cat" (4) does not fit into this group because the common cat is a domesticated animal whereas tigers, wolves, and lions are wild animals. Unfortunately, there is no way to make a firm determination because the set is not named.

### If the Name of the Set is Felines. . .

You might have chosen *wolf* as the inappropriate element because the other categories suggest that da Vinci was really trying to come up with taxonomic categories. The grouping of items in the set suggests that he has differentiated primate, feline, equine, and bovine classes. We know that a wolf is *Canis lupus*, a type of dog rather than a type of cat. A greater morphological difference exists between a wolf and a lion than exists between a cat and a lion. In addition, the series starts with *lion*, a word that could be used to name a set of types of felines or a term which names a class of wild animals.

### If the Name of the Set is Wild Animals. . .

Yet, if we thought that the class was made by a collection of wild animals, *wolf* would have more in common with *tiger* and *lion* than *cat* would, because a cat is typically a house pet. The common cat would not fit into this category of wild creatures. The difference between a wild animal and tame animal is to some extent a behavioral difference.

The point here is that for Set B we do not know what principle organizes the examples. The name of the set is an item in the set, so we do not receive much of a clue from the name. What was Leonardo da Vinci thinking when he put *wolf* in the same group with *cat* and *tiger*? We do not know. He does not tell us. He has not given us his ordering principle. He may even have blended two ordering principles in that series.

But we do understand da Vinci's reasoning for the other set, Set D. Da Vinci gives us his ordering principle in that one. Even though we might think it is strange to place a giraffe and a goat in the same group, we understand the reasoning that stands behind this grouping. He tells us: this group is made up of creatures with horns and without upper incisors. Set D has a specified ordering principle, but set B does not.

What is the ordering principle for Set C? For Set A?

Important Rule: Do not mix up separate ordering principles or sets of criteria.

# Applications

### Pre-Writing: Before you Write

Classification offers a useful tool in helping you get started when you start to write a paper, because organizing knowledge in the form of a paper requires that you make categories.

### Finding Your Topic

Your teacher has just assigned you to write a paper for the class and that is all you know at the moment. What will be the subject of your report? Where do you start? The substance of this discussion will direct you to (1) picking your subject and (2) deciding the treatment of the subject.

*Step One: Start with what you already know.*

First of all, you are in a science course, so you know that you should write a paper on a scientific subject. From Chapter 2, you recall that *any* topic may be treated technically. Although you are in a science class and any subject is open, you know that you will not be writing a poem, or a letter, or a stylistic analysis of a Shakespearean soliloquy. You know the genre, or kind, of writing you will engage in.

*Step Two: Pick your topic.*

Decide what you are already interested in or what you would like to learn more about. You are better off writing about something you like or are interested in than in feeling as if you have been forced into an odious and horrible task.

You do not have to have a lot of ideas about the topic—just pick something that you are generally interested in.

*Step Three: Brainstorm.*

Do not try to organize while brainstorming. Write down every bit of knowledge at your disposal. Write down what you do not know. Jot down notes. Write on your napkin at lunch, on a concert ticket, or on a deposit slip. Take time to make free associations.

Let us say you wanted to write an essay about water. That is the starting point. The topic of water.

What do you know about water?

Water is wet. It has hydrogen and oxygen. It melts and freezes at certain temperatures. It feels good to take a warm bath. Most of the earth is water. Most of your body is water. Water is everywhere. Is it everywhere?

### Step Four: Interrogate your notes.

After you write down everything you know, go back and ask yourself questions.

- *Water is wet.* What does wet mean? Can other things be wet? Some things are more wettable than others; why is that true? Does this have to do with the way that water is a universal solvent? Does it have to do with the surface tension of water? Are other things wet like water? Are all liquids wet? Is everything that is wet also a liquid—do wet solids or wet gases exist?
- *It has hydrogen and oxygen.* There are two hydrogens for one oxygen but the molecule is not balanced. What is the significance of the charged molecule? What is the nature of the hydrogen bond? How does the hydrogen bond relate to capillary action?
- *It melts and freezes at certain temperatures.* Is this always true? Under what conditions would it not be true? What if you tried to make ice in space? What is the ice crystal like? Why does water expand when it freezes? Does anything else make a crystal structure like water? Why? Would water freeze at the same temperature if there were something else in it?
- *It feels good to take a warm bath.* How do other substances respond to water? Why do your fingers shrivel in water? What happens if you put one hand in a pan of warm water and the other hand in a pan of cold water? What happens to your blood when the warmed blood from the warm water and cool blood from the cold water meet inside your body? Why can't humans be frozen and defrosted with injury? Why can sperm be frozen but persons cannot?
- *Most of the earth is water. Most of my body is water. Water is everywhere.* Is there water on the moon? Is there water inside of rock? Is there water in your bones? Is your body really filled with water? Isn't there some empty space inside your abdomen?

### Step Five: Make categories.

Some of the questions fall into the category of the chemistry of water: its structure in all states and the importance of that structure.

Another category might be how water reacts with other substances: with other water, with chemicals.

Another category might be the way that water is the biological basis of life and it is important for the way that electrolytes may be transmitted through the medium.

Wetness emerged as a category. Who would be interested in wetness? Farmers or entomologists who use pesticides know that some powders require the addition of a wetting agent in order to make the insecticide powder soluble.

*Step Six: Pick a category.*

Choose an approach that interests you: a biochemical approach, a sociological approach, a medical approach, an engineering approach, a physics approach. In other words, after you have derived the topic, choose an aspect of the topic to develop. Phrase your topic as a complete sentence.

*Step Seven: Turn your topic into a specific question.*

A question narrows your topic further.

To provide guidance in your literature search, turn the statement into a question. This can be a *what?* a *why?* a *how much?* or *how many?* or a *how?* question. The kind of question you ask will determine your orientation the subject. A *what* question deals with issues of substance and structure. A *why* question deals with cause-effect. A *how much/how many* question points you to quantitative analysis. A *how* question is a question of means; through what means does something occur?

If at this point, you have identified a topic and an approach to your topic, you will have classified your topic through a process of excluding other possible avenues to explore. The next step in this procedure is the trip to the library, but before you go, write down your question. By writing the question, you have defined an ordering principle. This principle will be the basis for your excluding certain texts from your search.

## The Library Search

Armed with your question, you will be relieved to discover that libraries use regular systems of classification. For your research, you may be interested in two categories of source material: old knowledge and new knowledge. When you look up your topic in books, you are dealing with old knowledge in part because books are often published as long as a year after they've been written. For new knowledge, go to journals, periodicals, or other monthly or quarterly publications. If you have access to computer information links, you may be able to get source material from the computer. If you need more clarification on your topic, start with the encyclo-pedia.

Using the computer search system or a card catalog, search for your topic as it is classified by subject, by author, or by title. Subject searches are sometimes organized by key words. If you cannot bring up the subject you need, check with the librarian or look around for a list of key words used by the computer. Match your classification to the library's system of classi-fication to generate a list of useful titles. Generate related words and search again.

For journals or articles, you will need to consult indexes unless you already have particular citations to look up.

### Choosing Your Texts

With your topic phrased as a question, you have an ordering principle to help you in your search. As you go through the lists of titles or articles, ask yourself if this book or this article will help to answer the question.

What if there are lot of books generally dealing with your question? Do you have to read all of them?

Obviously you want to strive for some kind of completeness, but it is better to gear your search toward those works which are most relevant to your concerns. So in your mind, define some more categories:

*Directly relevant* to my study
*Indirectly relevant*: some parts may pertain to my study
*Indeterminate relevance*: parts may be relevant but you would have to do a lot of work just to find that out.

Then look up those books and articles in the order of most relevant to least relevant.

### Classification in the Literature Search

As you read through the related work on your topic, categorize the works you read with respect to your topic. As your work, you will notice that your topic includes sub-topics. If you are engaging in research for an experiment, you readings will include will research on the methodology, on the back-ground of the subject, on related theoretical studies, on related experimental studies. Make these into categories.

Methods

> conventional method [1, 2]
> conventional method with significant variations [3, 4, 5, 6, 7]

Results
> analogous studies which corroborate my results: [A, B, C]
> analogous studies which qualify my results: [D, E]
> analogous studies which contradict my results: [F]
> unrelated studies which corroborate my results: [G, H, I]

While you may find a number of studies that corroborate your results, you may also find a contradictory study that limits the implications of your findings. Even though only one article exists to this effect, one article goes into the set of analogous studies which contradict results.

As you come to write up the literature section, you will integrate these notes by the organizing principle.

> While the contemporary view has noted that ... [A, B, C], this view has not been received unequivocally. Some research has indicated the opposite [F].

## Exercises

### 1. The Cownose Ray

The following paragraphs offer two descriptions of the same creature, a cownose ray. One description comes from a sixteenth-century book, Ambroise Paré's *On Monsters and Marvels*. The modern description comes from the *Audubon Society Field Guide to North American Fishes, Whales & Dolphins.*

Read each of these descriptions carefully. The Audubon guide is a field guide enabling the reader to identify a creature in the wild. Paré explains the ray is "a thing that never [before] been seen."

In a three to five page essay, analyze the connection between classification and identification. Consider how each work classifies the cownose ray, and on the basis of that classification, provides relevant detail. Make sure you discuss the graphics and the format of each selection.

An ell = four feet. Quioze is Chioggia in the Gulf of Venice.

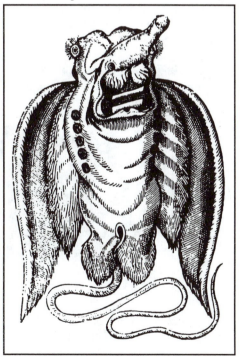

### Example 1: A Monstrous Flying Fish

Between Venice and Ravenna, one league [two miles] above Quioze in the sea of the Venetians in the year 1550, a flying fish was caught that was frightening and gave marvel to see, [being] four feet more in length and twice as much in width from one tip to the other of its wings, and a good square foot in thickness. Its head was wondrously thick, having two eyes, one on top of the other, in a line; two large ears and two mouths; its snout was very fleshy, green in color; its wings were double; on its throat it had five holes in the fashion of a Lamprey; its tail was an ell long, on top of which were two little wings. It was brought quite alive to said city Quioze, and presented to the lords of latter, as a thing that had never [before] been seen.

### Example 2: The Cownose Ray*

(*Rhinoptera bonasus*)
Description:

To 3' (91) cm. wide. Disc about 1 1/2 times wider than long, front edges nearly straight, posterior edges of disc concave, outer corners falcate. Brownish, above, whitish or yellow-whitish below. Front of head moderately concave, *subrostral fin deeply notched in the middle, forming 2 lobes paired at back*, head and subrostral lobes form shape resembling a cow's nose. Tooth plates usually of 7 series of teeth. Tail spine immediately behind dorsal fin; skin smooth.

Habitat:
Primarily bottom of shallow bays and inshore shelf.
Range: From New England to Brazil. Caribbean and the Gulf of Mexico.

Comments:
Cownose rays feed primarily on hard-shelled mollusks that they crush with their powerful tooth plates. They sometimes occur in bays during the summer and disappear in the winter. Their stinging spines make them potentially dangerous.

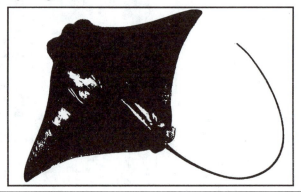

---

* *The Audubon Society Field Guide to North American Fishes, Whales, and Dolphins* (New York: Alfred A. Knopf, 1983), p. 360).

## 2. Virus: Living or Not?

In the Reading Selections, turn to M. W. Beijerinck's essay, "A *Contagium vivum fluidium* as the Cause of the Mosaic Diseases of Tobacco Leaves," and analyze his efforts to assign a classification.

## 3. Hippocrates on Epidemics

Turn to Hippocrates, "Epidemics," in the Reading Selections. What do the sufferers of *causus* have in common besides the illness? Does Hippocrates use more than one ordering principle? Does he mix ordering principles?

## 4. Anomalies

The following abstract describes the nature and significance of anomalies. Identify a past or present anomaly in your field, and explain the significance it holds for past or current theories.

An anomaly in science is an observed fact that is difficult to explain in terms of the existing conceptual framework. Anomalies often point to the inadequacy of the current theory and herald a new one. It is argued here that certain scientific anomalies are recognized as anomalies only after they are given compelling explanations within a new conceptual framework. Such a "retrorecognition" phenomenon reveals not only a significant feature of the process of scientific discovery but also an important aspect of human psychology. —Alan Lightman and Owen Gingerich, "When Do Anomalies Begin?" *Science* 255 (7 Feb. 1992): 690–94.

## 5. Territorial Differences

Identify an issue in the history of science of which has been "claimed" as the territory of two different disciplines. (How do paleontologists and biologists define "species"?) Discuss how each field has treated that subject, and speculate on the possibility of finding a common ground.

## 6. A Flying Machine

When designing his flying machine, Leonardo da Vinci turned to birds as his models for flight. His interest in them was determined by his research into flying machines. When he classified bird, he did not employ a taxonomic categorization. He wrote

A bird is an instrument working according to mathematical law, which instrument is in the capacity of man to reproduce with all its movements but with as much strength, though it is only deficient in the power of maintaining equilibrium.[4]

Analyze da Vinci's classification.

# References

*Audubon Society Field Guide to North American Fishes, Whales & Dolphins*. New York: Alfred A. Knopf, 1983.

Hayakawa, S. I. "Classification," in *Language in Thought and Action*. Harcourt, Brace, & Co., 1941, 1949.

Paré, Ambroise. *On Monsters and Marvels*. Translated by Janis Pallister. University of Chicago Press, 1982.

## Notes

1.  Ferdinand Schoeman, "AIDS and Privacy," in *AIDS and Ethics*, edited by Frederick G. Reamer. (New York: Columbia University Press, 1991), p. 244.
2.  S. I. Hayakawa, "Classification," in *Language in Thought and Action* (New York: Harcourt, Brace, & Co., 1941, 1949), pp. 215–17.
3.  *The Notebooks of Leonardo Da Vinci*, ed. Pamela Taylor (New York, London, Scarborough: New American Library, 1960), pp. 115–16.
4.  *The Notebooks of Leonardo Da Vinci*, pp. 101–2.

# 11

## Comparison/Contrast

Comparison refers to likeness; contrast refers to differences. Comparison/contrast refers to a method of describing the differences and similarities between things, events, phenomena.

There is no contrast without comparison.

There is no comparison without contrast.

While comparison and contrast are theoretically implicated in each other, it is not unusual to find that an essay or paper will be slanted more toward one aspect than the other. A paper which proposes to show a correlation between two items will emphasize the similarity which exists despite differences. A paper which attempts to demonstrate the novelty or variation will present the differences which exist despite the similarities. The proportion of comparison to contrast is determined by your purpose in writing. It directs the organization of your points by imposing a hierarchy of priorities.

### To Write a Comparison/Contrast

To write a comparison/contrast, you must preserve an orderly arrangement and you must direct the course of your comparison with a purpose.

#### 1. Parallelism

What you say for A, you must say for B. Your comparison/contrast will be unbalanced if you discuss three points for one term and one point for another.

#### 2. Order

Use the same *sequence of points* for discussing every topic.

Note: Your thesis or statement of purpose should identify both the points you will present and the *order* of their presentation.

### 3. Audience and Purpose

Be aware of your purpose and your audience. Do not discuss details randomly. Choose a pattern of organization—Divided or Alternating (see below)—that best supports your point.

*Purpose*: Comparison/contrast is a means to an end not an end in itself. To illustrate the importance of having a purpose, we will consider the obverse by imagining a circumstance for comparison/contrast which is purposeless. For the sake of argument, we will compare/contrast apples and oranges:

Apples are round; oranges are round.

Apples are fruit; oranges are fruit.

But apples are red, whereas oranges are orange.

Both have seeds.

This is a meaningless exercise because the c/c neither leads to nor supports any point that is greater than the obvious one that apples differ from oranges. Because c/c is a strategy rather than a purpose in itself, you should use comparison/contrast as *a means to an end*. You should, therefore, avoid a statement of purpose or a thesis which states *Even though they are alike, they are different* OR *Even though they are different, they are alike*. Your statement of the issue should address the *significance* of the likeness or difference.

## Two Patterns of Comparison/Contrast: Divided and Alternating

Below you will find a passage that compares and contrasts bone and shell. This passage, both in its adapted form and in the original,[1] argues for a linkage between vertebrate and invertebrate lines of evolution. Before you read the passage in its entirety, we will outline the writing task and demonstrate the two ways that this comparison/contrast could be articulated through the Alternating and the Divided patterns of organization.

While both bone and shell are types of hard tissue, bone forms the internal system of support for vertebrates whereas shell forms an external support system for invertebrates. Yet despite these differences, bone and shell are generated in similar ways: bone, like shell, is produced by the mineralization of a matrix laid down by specialized cells.

*Bone* and *shell* are the two terms to be contrasted and compared. What parallel points may be discussed for each of them? Because both bone and shell may be discussed in terms of their similar substance, their chemical composition offers a point of contact. The process of *formation* offers another point of contact between these two substances, and the *function* of bone/shell in the structure of the organism provides a third contact point .

A discussion organized in the Divided Pattern would divide the body of the discussion into two sections, one on bone (substance, process of formation, function) and one on shell (substance, process of formation, function), in addition to the introductory and concluding paragraphs. In other words, the discussion is divided to treat one topic fully before it goes on to discuss the second fully, in the same order.

The Alternating Pattern would generate three sections: on substance (bone and shell), on the process of formation ( bone and shell), and on the function (bone and shell).

| Divided Pattern | Alternating Pattern |
|---|---|
| I. Introduction | I. Introduction |
| II. **Bone** | II. Substance |
|    A. Substance |    A. **Bone** |
|    B. Process of Formation |    B. **Shell** |
|    C. Function | |
| III. **Shell** | III. Process of Formation |
|    A. Substance |    A. **Bone** |
|    B. Process of Formation |    B. **Shell** |
|    C. Function | |
| | IV. Function |
| |    A. **Bone** |
| |    B. **Shell** |
| IV. Conclusion | V. Conclusion |

Pattern A, the *Divided Pattern*, is so-called because it divides the discussion into two parts. It is also called the *block form* because the two separate sections make up two large blocks of writing. All the criteria for each block—substance, process of formation—are grouped together, not juxtaposed. Because the divided pattern treats one topic at a time, it softens the contrast between the elements under discussion.

Possible Limitation: One limitation of the divided pattern is that it sometimes directs the writer into writing two separate illustration paragraphs—or two illustration papers. This problem is the most frequent error in using the divided pattern.

*What's wrong with writing two separate illustration essays?*

Writing two illustration essays undercuts the purpose of writing a comparison/contrast because the connection between the compared elements becomes lost. Writing two illustration essays also places the writer in the position of doing more work for less content.

To prevent yourself from writing two separate illustration essays, discuss the two elements—bone and shell—with reference to each other. Even though you do not directly juxtapose discussions of the two elements in one paragraph, maintain reciprocal reference between the two items by the following: (1) Refer to details of paragraph 2 when you write paragraph 3, and use words which show comparison/contrast relationships: "like," "unlike," "similarly," "in contrast." (2) Juxtapose the two items in your conclusion; you would be using the alternating style here.

The *Alternating Pattern* juxtaposes the topics. When you place two topics or points side by side, you make it easier for the reader to assess them with respect to each other. The Alternating Pattern enhances the contrast between the compared elements. But don't take for granted that placing two elements side by side will automatically explain your point. Help the reader to understand your point by explaining what you mean to prove through the comparison/contrast. The key rule here is *Explain the obvious*. Do not take for granted that the comparison is self-evident.

### Divided/Block Pattern: Examples

The following selection, organized in the block form, compares and contrasts bone and shell. The first paragraph discusses bone; the second paragraph discusses shell by its likeness to bone.

> (1) Bone is a distinctive hard tissue peculiar to vertebrates. (2) It is produced by the mineralization of a matrix elaborated by specialized cells. (3) The cells active in the formation of hard tissue are drawn from the ectodermal and mesodermal layers of the embryo. (4) These matrices secrete fibers of polysaccharide or polymerized protein. (5) Bone is unique in its microscopic structure and in the particular characteristics of its cell, matrix, and mineral content.
>
> (6) Though laid down by invertebrates, shell is *likewise* a form of hard tissue. (7) Mollusk shells and the hard parts of lowly metazoa are produced by mineralization of a matrix produced by mesodermal or ectodermal cells. (8) The matrices they secrete contain fibers of polysaccharide or polymerized protein. (9) For instance, brachiopods, or "lamp shells," have been laying down calcium as hydroxyapatite, the phosphate rich salt found in bone, since early Ordovician times

### Alternating Pattern: Example

In the following selection, the comparison/contrast between bone and shell is given in the alternating pattern.

(1) Bone has much in common with the hard tissues produced by other living things. (2) Bone is like arthropod exoskeletons, mollusk shells, and the hard parts of even more lowly metazoa because they are produced by the mineralization of a matrix or ground substance elaborated by specialized cells. (3) Both vertebrate and invertebrate cells active in the formation of hard tissues are drawn from both the ectodermal and the mesodermal layers of the embryo. (4) Their matrices are alike in that they almost always contain fibers of polysaccharide or polymerized protein. (5) The minerals deposited in the ground substance of the hard tissues of most animals generally consist of calcium carbonate or phosphate in an amorphous form or as crystals of various sizes. (6) Even though bone is unique in its microscopic structure and in the particular combination of characteristics of its cell, matrix, and mineral content, its constituents exist in other animals. (7) For instance, certain brachiopods, or "lamp shells," lay down calcium as hydroxyapatite, and this salt is found in bone. (8) Many sponges have produced collagenous fibers like those in vertebrate osteoblast matrices. (9) Therefore, bone resembles shell in the process of its formation.

## Using Comparison/Contrast to an Effective Purpose

In the following passage, Barbara Stahl responds to a common view about evolution which argues the following: Some taxonomists believed that emergence of bone signaled an innovation in evolution. The consequences of this innovation are significant because bone offered an evolutionary advantage which led to the development of vertebrates when bony fishes superseded cartilaginous fishes. In other words, some taxonomists believe that bone, an "innovation," led to the evolution of a "new" kind of animal, vertebrates.

But Barbara J. Stahl takes a different view from this traditional understanding of the place of bone in vertebrate evolution. In arguing that, while the substance of bone was an innovation that led to evolution of vertebrates, the process by which bone is made is not new. In fact, she maintains, vertebrates make bone through a process akin to the way that mollusks make shells or crabs make exoskeletons. This difference implies that a stronger connection exists between invertebrates and vertebrates than had hitherto been thought.

Stahl's central point is the assertion of a kinship between vertebrate and invertebrate evolutionary lines; that assertion is carried through a c/c between bone and shell. In the following passage, look for the way that the writer keeps in focus the likeness between bone and shell.

(1) Although bone is a distinctive tissue peculiar to vertebrates and has not yet been traced to a particular precursor tissue at the invertebrate level, research has shown that it has much in common with hard tissues produced by living things other than vertebrates.

(2) Bone, like arthropod exoskeletons, mollusk shells, and the hard parts of even more lowly metazoa is produced by the mineralization of a matrix or ground substance elaborated by specialized cells. (3) Among vertebrates and invertebrates, cells active in the formation of hard tissues are drawn from both the ectodermal and the mesodermal layers of the embryo. (4)The matrices they secrete are alike in that they almost always contain fibers of polysaccharide or polymerized protein. (5) The minerals deposited in the ground substance of the hard tissues of most animals consist generally of calcium carbonate or phosphate in an amorphous form or as crystals of various sizes. (6) Bone is unique in its microscopic structure and in the particular combination of characteristics of its cell, matrix, and mineral content, but the constituents of the vertebrate tissue existed in other animals before backboned forms appeared. (7) For instance, certain brachiopods, or "lamp shells," have been laying down calcium as hydroxyapatite, the phosphate-rich salt found in bone, since early Ordovician times, and many sponges of great antiquity have produced collagenous fibers like those which appear in the matrix formed by vertebrate osteoblasts. (8) Bone, it seems, did not originate in early vertebrates as a tissue completely new in every sense but simply represented a new combination of materials assembled in an already old and well-established way.[2]

## Further Discussion of the Example

The Stahl excerpt may be described, from a writing a point of view, in the following way:

| | |
|---|---|
| Audience | An educated but unspecialized lay audience |
| General Organizational Mode | (1) a *comparison/contrast* between bone and hard tissue |
| Refinement | (2) focusing on the *similarities* |
| Purpose | (3) to show a stronger connection between vertebrate and invertebrate lines |
| How Contrast fits in | (4) the general view already assumes that vertebrates and invertebrates are more different than alike. |

Stahl's paragraph argues for similarities because the differences are already taken for granted. The common view holds that the bone in your body is different from the pearly substance you see lining a sea shell. Bone is what humans or dogs and cats have; shells are what snails have. So Stahl need not elaborate the contrast because the general view already holds that vertebrates and invertebrates are more different than alike. *The contrast is built into the context,* and it is subordinated to her argument which is based on similarities.

If you look again, you will see that all the comparisons are directed to dealing with that difference. The contrasts are subordinated grammatically, organizationally, and conceptually to resemblances.

## Exercises

### Analysis of the Comparison/Contrast Form

#### 1. Audience Analysis
Explain how the targeted audience governs the writer's organization. How would Stahl's selection be different if the audience were a general audience?

#### 2. The Archaeopteryx
The Archaeopteryx was a prehistoric bird that displayed both reptilian and bird traits. In order to classify the Archaeopteryx, the author compares/contrasts the pattern of bone growth in birds and reptiles with that of the Archaeopteryx. Analyze the use of comparison/contrast in this passage.

> Part of the problem with classifying *Archaeopteryx* stems from ignorance of whether their growth pattern, as recorded in their bones, was more reptilian than birdlike. Reptiles grow throughout their lives . . . ; conversely, birds quickly attain a characteristic adult size. In reptiles, the center of growth are the shafts in their hollow bones, whereas growth in young birds take place at the bones thick cartilaginous ends, called epiphyses. During the final stage of a bird's growth, its epiphyses turn from cartilage into bone, leaving a scar that disappears when the bird matures.
>
> None of the *Archaeopteryx* specimens shows any such scars on its hollow bones. If the growth of these animals were birdlike, then the various specimens might indeed represent different species. On the other hand, if they had reptilian growth patterns—an assumption supported by the predominantly reptilian traits of the skeletons—then the specimens could clearly be members of the same species, different in size and age. Recent studies . . . support the view that the six specimens of *Archaeopteryx* represent different stages of growth of a single species.[3]

#### 3. Models of Disease
a. Turn to the Reading Selections, and after reading Hippocrates, "Epidemics," and Thucydides, "Plague of Athens," compare and contrast the models of disease which underlay their views.

b. Introduce Fracastoro's "Germ Theory of Disease" into the discussion to position his work with respect to the other two writers.

#### 4. AIDS and Syphilis
In each of the following, analyze the writer's use of the comparison/contrast

mode in discussing AIDS and syphilis. Answer the questions that follow the readings.

Example 1

AIDS, obviously, is a disease considerably different from syphilis. Caused by a human retrovirus, human immunodeficiency virus (HIV), it attacks the immune system, leaving affected individuals vulnerable to a host of other infectious agents. In addition to being transmitted sexually, HIV can be transmitted through blood transfusions as well, thus making intravenous drug users who share needles a principal risk group for infection. Unlike the statistics for syphilis, in AIDS, principal morbidity and mortality occur among young people; the most serious consequences of syphilis usually come in the late stage of the disease, among older individuals. Finally and most significantly, at this moment there are no curative treatments for AIDS and no means of rendering infected individuals noninfectious. Despite these important differences, however, the history of syphilis presents a series of striking parallels to the many problems raised by AIDS.

AIDS, like syphilis in the past, engenders powerful social conflicts about the meaning, nature, and risks of sexuality; the nature and role of the state in protecting and promoting public health; the significance of individual rights in regard to communal good; and the nature of the doctor-patient relationship and social responsibility. The analogs that AIDS poses to this brief history of syphilis are striking: the pervasive fear of contagion, concerns about casual transmission, the stigmatization of victims, and the conflicts between public health and civil liberties. The response to AIDS will be a function of our own time, our own culture, and our own science. The importance of the history of syphilis is that it reminds us of that range of forces that influence disease, health, and social policy.[4]

Example 2

One epidemic that is hauntingly similar to AIDS is syphilis. The parallels are striking. At the time of the syphilis epidemic, scientific investigation of this insidious disease was at the leading edge of medicine and microbiology, as is the current situation with AIDS. The issues raised included public health policy and civil liberties, as in the AIDS epidemic. And finally, because it is a sexually transmitted disease, syphilis patients were highly stigmatized.

Why did it take so long to control this epidemic? Like AIDS, syphilis can be a long-term and variable disease, with phases in which no symptoms are apparent. Unfortunately, untreated syphilis eventually leads to death. More important—at time syphilis was perceived as a social problem—hence the reference to it as a social disease. Many blamed the disease on a breakdown of social values, and promoted the view that the sexual ethic in which all sex was marital and monogamous would make it impossible to acquire the disease. The initial public health policies to control this epidemic were based on these views. Abstinence from marital sexual conduct was encouraged and prostitution was repressed since prostitutes were blamed as the major source of infection of otherwise monogamous males. Immigrants were also blamed for bringing the disease from abroad, even though epidemiological data did not support this view. As many as 20,000 prostitutes were quarantined or jailed during World War I. In addition, the Army discouraged the availability of condoms for fear that they might encourage soldiers to engage in extramarital sex. There were also campaigns to stigmatize soldiers who became infected with syphilis by giving them dishonorable discharges. These policies were not based on epidemiological evidence, and they failed to control the epidemic, which actually grew during this period.

It was not until the 1930s that the surgeon general of the United States, Thomas Parren proposed major changes in the public health approaches to control the syphilis epidemic. These policy changes were ultimately successful but required substantial funding from Congress.[5]

(1) Do the passages emphasize the likeness or the difference between AIDS and syphilis?
(2) Are both passages directed to the same kind of audience? To different audiences?
(3) Explain the different emphasis each passage gives to the topic.

# Using Comparison/Contrast

## Essay Examination

Comparison/contrast can be used most effectively by combining it with other modes of writing like definition, classification, and illustration. For instance, if you were asked to answer an essay question like "Was Einstein a genius?" your approach would be to give a definition of genius and then c/c Einstein with the definition. (See the section on *Definition* for instructions on how to write a sentence definition.) You could use this pattern to answer any of the following questions.

(1) Is Plutonium a heavy metal?
   Define heavy metal; then c/c the attributes of Plutonium with those of heavy metals.
(2) Was Hamlet a tragic hero?
   Define tragic hero; c/c Hamlet with the traits of a tragic hero.
(3) Assess the success of the Marshall Plan.
   Define the Marshall plan; c/c the Plan's stated aims with its actual results.
(4) Is light a particle or a wave?
   C/c the behavior of light wave/particle behavior.
(5) Is a virus a living creature?
   Define living/non-living; c/c the attributes of a virus with those of living/ non-living creatures.

These questions do not require an absolute resolution, nor must you suppress the points which contradict your point of view. Sometimes your teachers want you to discuss the points of conflicting evidence in order for you to demonstrate that you know the complexities of the issue. You can use c/c either to come up with a decision or to open up an explanation as to why an answer is difficult to come by. Sometimes your answer to a question like (4) or (5) could be "both" or "neither," but to arrive at that conclusion, you must demonstrate your logical process.

**Statement of Purpose**

In a statement of purpose, such as an application to graduate school, you explain your intentions. A statement of purpose is usually given in a particular communication context: you write one because you want the person who reads it to act upon your writing in a certain way. You may be writing a statement of purpose because you want to be accepted to graduate school, to medical school, or to a particular academic program. You may also be requesting funds, equipment, faculty assistance or direction, computer or library support. Your goal is to have your request granted. (Note: The writing strategy which follows may also be adapted to a job application.)

A request for a statement of purpose may contain two different questions:

(1) *Why do you want to do what you propose in general?* In other words, why do you want to go to medical school, to graduate school, etc.? If you perceive the question to be one that elicits an explanation of your general motivation, one appropriate response may be personal narrative: Ever since my mother's life was saved by open-heart surgery, I wanted to go to medical school.

(2) The statement of purpose may be construed in another way. *Why do you want to do what you propose HERE? Why are you asking us?*

If you are applying to a school or a program, you need to demonstrate that you have needs or desires which that program can satisfy. To strengthen your argument, you might add that that program can satisfy your goals in a way no other place can. Perhaps a faculty person in that program or field does research in your preferred field. Mention your interest in that person's work. Suppose that Stanford had the only super-collider which would enable you to test your theories about the behavior of certain particles: bring up that point.

To organize your writing strategy, (1) define clearly to yourself what you want; (2) understand yourself enough to know why you want this; (3) research the institution for information about its program, its resources, and its faculty. Such research enables you to explain how you suit the school and how the school suits you.

Understanding your own motivation is important because you need to show that you possess mature judgment. It would probably show immature judgment if you explained that you were interested in a school because it had a cute campus, was close to the beach, offered a rich stipend, or ensured that all of its graduates got high-paying jobs. Keep in mind that you want them to want you. After all, *comparing* and *matching* connote reciprocity.

What would make any school, program, job want you? You need to demonstrate (1) interest, (2) ability, and (3) logical necessity.

## Interest

You can show professional interest or personal interest or both. By researching the school, you show serious professional interest. As it is appropriate, explain how a particular event or the influence of a particular person moved you to choose this career. Or, perhaps, this field was something that you were always interested in, and indeed, everything you have done in your life up to this day reflects that interest. There are many ways to show this: past interest, current interest, particular influences. Above all be specific.

## Ability

To a reader or a reviewer, past performance implies future performance. To show ability, explain what relevant work you have already done. Mention courses you have taken. State your strengths as they have been demonstrated by various tests. Explain particular accomplishments achieved by way of personal projects.

## Logical Necessity

To argue from logical necessity, present the argument that this school best serves your interest. Your writing should convey that your goals could be best served by no other alternative. If Stanford had the only super-collider that would permit you to conduct your research, indicate that your work imposes access to this instrument as a necessity. Note that logical necessity differs from emotional or financial necessity. Keep in mind that no institution owes you a place simply because you need it.

By using an organized format to show how you match the school, job, program of interest, you show that (1) you are an organized person who has focus and initiative such that you have done your homework on the institution, and (2) you know how to organize your writing effectively. Because public institutions most frequently make first contact with you through your writing, *who you are* and *how/what you write* will be intertwined. A person having to make a decision on your statement of purpose should come to the conclusion that you, as a person, are as organized, clear, and focused as your writing, and therefore, both (1) and (2) add up to (3): You create a credible image of yourself through writing.

## Proposals/Recommendations: C/C Real/Ideal

Comparison/contrast is a most effective tool when you use it to make recommendations. A recommendation can be framed as the resolution between *what should happen* (ideal) and *what did happen* (real). What should happen can be predicted on the basis of a theoretical or mathematical model; this happening can be rephrased as a hypothesis or as a predicted or usual result arrived at under specified circumstances. What did happen can be understood to be experimental results, data, the actual outcome of an experiment.

To make a recommendation or to propose a course of action, explain how a certain course of action brings the ideal case in line with the real. This arrangement can be geared to proposals by explaining that, since the experimental outcome differed from the theoretical predictions, further research is needed to supply the explanation which accounts for the discrepancy.

## The Scientific Paper

Most specialized technical scenarios can be found in science writing. Often a formal report about a scientific matter will compare and contrast

- real and ideal
- predicted results to actual results
- theory with experimental evidence
- your findings with other people's findings
- your methods with other people's methods
- your interests, ability and need to what a job, school, program offers

## Discussion Section: Why did mine turn green?

When you conduct experiments in a lab with other people around you, you know how common it is for you to look and see if other persons in the lab got the same results as you did. If you are in a lab and you notice that the liquid in your flask turned green while everyone else's turned blue, you have a sign that you did something differently. The discussion section of you write-up offers the opportunity to explain the discrepancy. Maybe you discovered a new species of bacteria. Maybe your reagent was contaminated. Whatever the reason, explain it. You don't have to know all the answers.

There are several levels from which to draw for the explanation as to why yours turned green.

### 1. Procedural Level

You decide on the way to conduct your experiment, or you have conducted it in a certain way. You might find in the literature search or by talking with people in your lab that someone conducted this experiment in the same way or that someone conducted a similar study in a different way. Their experiment was the same or similar to yours but it differed only in terms of

- special equipment or conditions
- qualitatively different new technology
- different but related substances
- other animal or plant species

Their study could be different from yours in more than one of these aspects, but keep in mind that if their work differed from your work in too many of these aspects, you would have a completely different experiment.

### 2. Theoretical Level

How does your work fit into the field in term of its theoretical model? Does an alternative explanation account for different results under similar conditions? Is there a better explanation that could explain similar results under similar conditions?

Presumably, you started with a theory or a hypothesis about what will happen or what the results your study will eventually show. Did other people in your field have the same idea, use the same model? Did other people in related fields have the same idea? Is there a theoretical explanation for why you arrived at one set of results and someone else arrived at a different set of results?

In other words, compare and contrast different theories to account for a discrepancy in result

### 3. Conceptual Level

Between the theoretical explanation you use to frame your experiment and the actual practice of that experiment is an intermediate level which might be called the conceptual level. The conceptual level does not necessarily have recourse to a theoretical model, although a theoretical model could be used to help explain things. We use the conceptual level as a mode of thinking to direct our investigation. For instance, when you take a subject and compare what you know about it to what you do not know, you are taking a conceptual approach. By comparing the known to the unknown, certain conclusions can be derived; one conclusion might be that further study is needed, or that you need to repeat your experiment to figure out why yours turned green.

### Wording: How do you say it?

Certain expressions are useful for discussing similarities and differences in science writing. The following offers a few of the most common expressions. These are not the *only* ways to say things; they are just *some* of the ways.

The strongest statement of concord (similarity, agreement) or of discrepancy (difference, disagreement) comes through the use of verbs.

| predicted experimental mathematical empirical | results | confirmed matched supported verified agreed with show good agreement with concur with coincide with | our/their | predictions hypothesis theory |

In addition to the "not" forms of the agreement verbs (X does not support Y), you can also use single verbs to show discrepancy: *refute, disagree with, contradict, vary from, diverge from. Like, unlike, similarly, on the other hand, in a contrary manner, conversely, resembling* also signal the relationship.

## References

Brandt, Allan M. "The Syphilis Epidemic and Its Relation to AIDS." *Science* 239 (22 Jan. 1988).

Fan, H., R. F. Conner, and L. P. Villarreal. *The Biology of AIDS*, 3d edition. Jones and Bartlett Publishers, 1994.

Stahl, Barbara. *Vertebrate History: Problems in Evolution*. Dover Publications, 1975, 1985.

Wellnhofer, Peter. "*Archaeopteryx*," *Scientific American* (May 1990).

## Notes

1. This passage, found in its complete form later in the chapter, is drawn from Barbara J. Stahl's *Vertebrate History: Problems in Evolution* (New York: Dover Publications, 1975, 1985), pp. 55–56.
2. Ibid.
3. Peter Wellnhofer, "*Archaeopteryx*," *Scientific American* 74 (May 1990): 70–77.
4. Allan M. Brandt, "The Syphilis Epidemic and Its Relation to AIDS," *Science* 239 (22 Jan. 1988): 375–80.
5. Hung Fan, Ross F. Conner, Luis P. Villarreal, *The Biology of AIDS*, 3d ed. (Boston and London: Jones and Bartlett Publishers, 1994), pp. 20–21.

# 12

# Process

Chapter 8, Description, explained how to describe something as an object in space. In this chapter you will learn how to describe something which happens in time. Previously, in describing an *object*, you learned how to break an object into parts and then to explain how the parts fit together spatially, functionally (in order of use), or chronologically (in order of assembly). In this section, you will emphasize a *process*—how something happens in time. Like the description of the object, a description of a process requires you to break the events of a time sequence into parts and then to structure the parts in the way most relevant to your purpose and to your audience.

## Three Ways to Describe a Process

When you describe a process you tell *how* something happened. You can classify the question of *how* into 3 different modes: (1) how *I* did it, (2) how *you* can do it, (3) how *it* happens.

### 1. Process Narrative: How I Did It

Narrative refers to relating successive events which have time in common. This mode of writing focuses on the *agent*: what I did. Thus process narrative is writer-oriented. For instance, I explain

    how **I** made bread;

    how **I** stained a slide;

    how **I** operated a type of equipment;

    how **I** performed an analysis.

    A lab report or the methods section in a formal paper are the most frequent types of process narrative you will employ.

To explain how you did something, divide your process into a list of major steps and sub-steps as they apply.

This form may take the usual three-part essay development: introduction, body, and conclusion. The introduction tells what you did, how, when, why, and where you did it. The body deals with how you did it; the list of steps is given in the body. The concluding section tells what the results were or what was achieved.

Your title should be clear and comprehensive. Your steps should be logically ordered and phrased in parallel terms. Your explanation should adopt the level of technicality appropriate to your audience. Remember the rule: the less your reader knows, the more you need to explain—by way of written text or graphics.

## 2. Set of Instructions: How You Do It

The emphasis in a set of instructions falls on the *reader*: You tell someone else how to do something. A recipe, a computer operator's manual, a set of instructions on how to assemble a bike fall into this category of "you" focused writing. For instance, you explain to another person

how *you* make bread;

how *you* stain a slide;

how *you* operate a type of equipment;

how *you* perform an analysis.

Your knowledge of your audience is especially important here. Because another person will actually follow your instructions, instructions must be absolutely clear to the reader.

Let us suppose that you are in a two-seater airplane on your way to Catalina, and the pilot has a heart attack. You are virtually alone in the airplane. Because you know to use the radio, you call in to the local tower and ask for instructions on how to land.

If the air traffic controller tells you to keep the plane level for twenty miles, then descend over a local air strip, you are still in trouble because you do not know how to keep the nose level, how to descend, ascend, steer, lower the landing gear, and brake. In very precise detail, the air traffic controller must tell you not only the steps but also the precise way to perform each step and the location of the instrumentation needed to perform the step.

Fortunately, the air traffic controller has given you very clear steps. He told you where to locate the controls and how to operate them. He also explained the importance of pertinent gauges: the altimeter shows how high you are. If the altimeter shows 0 feet altitude, you will be flying at sea level—most runways are *above* sea level. He calls your attention to the gas gauge, the oil pressure gauge, the warning lights. He gives you a general overview of the process.

You start to relax. Flying, you discover, is just like driving.

Now the air traffic controller tells you: "Begin your descent by pushing in on the wheel slowly."

You do this. Suddenly the voice over the radio asks "Have you started your descent?"

"Yes," you answer.

"What did you do about that big water tower?"

"I went around it," you say.

"I should have told you sooner," the voice says, "we lose more small craft to that water tower."

Fortunately, you saw that giant water tower and dodged it, but you would have appreciated knowing it was coming. You dodge it. You land safely. But you did not like that surprise.

The following remarks summarize the main points for writing a set of instructions:

1. Know your audience. Instructions are written at their level.
2. Use the imperative. Instructions are directions TO a you: *you* do this, *you* do that. The imperative is the one grammatical case in English where the subject of the sentence may be omitted because it is implied: (You) turn off the engine. (You) get out through the door.
3. Explain the function or purpose of apparatus or procedure. This explanation must be complete. List *all* necessary equipment *in order* of use. Give a summary of major steps and principles of operation, as they apply, before you explain the separate steps.
4. Divide the instructions into logical steps and explain each step in order.
   Each step should deal with one level of operation. Maintaining altitude, Steering, Lowering the Landing Gear, Beginning Descent, Touch Down, Braking, and Turning Off the Engine would be separate steps in this procedure.
5. Divide steps into substeps as necessary.
   Touch down has five substeps:
   (a) Leveling the nose
   (b) Trimming the flaps to stabilize the wings
   (c) Contacting with ground
   (d) Readjusting nose and wings
   (e) Decelerating
   Note: Deceleration is a different step than braking because "not moving forward" is a separate act from "stopping." Deceleration means you do not apply gas to move forward. Braking means that you do something particular to stop.
6. Give *all the steps* necessary to the procedure per the experience of the operator.

You would be in trouble if you managed to land safely but did not know how to stop the plane once it was on the ground. You also need to know how to turn it off and get out because neither you nor the ground crew want to step into a turning propeller.

7. Give warnings, notes, danger signs and preconditions BEFORE the explanatory step.

For instance, the air traffic controller should have told you a water tower stood in the path of your descent *before* he told you how to make your descent. If you have used a cook book, you will know that frequently the first instruction for baking is *Preheat the oven* even though the last thing you do is put the food in the oven.

8. When you write instructions, phrase the steps in grammatically parallel terms. "Braking" and "Leveling the Nose" are all phrased in the "ing" forms.

### 3. Process Analysis: How It Happens

The emphasis on this writing falls on the *subject*: what happened. Process narrative is subject-oriented:

how bread is made;

how a stain takes in a cell;

how an analysis is performed.

Process analysis deals with relationships in time and how things are made. Because it concerns time, it is cited in the chapter on description; however, its place in this chapter is formal. Process analysis extends conceptually into the realm of causal relations.

# Causality

Causality refers to the relationship between causes and effects, i.e., to the way that one object, phenomenon, or state gives rise to, prompts, creates another. Other verbs which designate *cause* include:

originate, give rise to, bring to pass, make, produce, create, develop, beget, contrive, effect, effectuate, engender, foment, generate, incur, induce, precipitate, prompt, provoke, inspire, trigger, evoke, elicit.

Note: before you use a synonym for *cause*, be sensitive to the nuances of the word. These words range in meaning from "directly producing an effect" to "calling an effect into being."

Cause-effect proposes a relation between two elements, such that no B is produced in the absence of A; and that the presence of A and A alone produces B.

There are two kinds of writing tasks associated with cause-effect: One starts with the effect and works backwards to causes; the other begins with cause and generates effects from it. If you are assigned a causal analysis paper, you must identify your starting point as either cause or effect.

In general, an essay which analyzes cause gives students the most problem, in part because it calls for the most abstract thinking. When you analyze cause, you try to explain why something happened. When you analyze an effect, you explain the consequences or results of a certain action.

Cause: Questions of cause may be *why* or *how* questions. What causes cancer? What causes AIDS? What causes epidemics? What causes measles? What causes electricity or storm? Why do students get good grades, bad grades? Does TV violence cause social violence? Why can a frog reproduce parthogenetically?

Do not confuse remote cause with proximate cause. See below.

Effect: It is usually easier to analyze and explain effects because you work backwards from a known result. Analysis of effect also deals with *how* and *why* questions. When your chemistry instructor asks you to explain what happens when mercury chloride is mixed with mercury metal or when you write the Discussion section of your formal report, you analyze effects.

The Discussion section of formal papers deals with a discussion of effects and how they came about. The Discussion deals with whether these were expected or predicted results (per your hypothesis or compared with other work in the field) or accounts for discrepancies between expected results and actual results. A research question may be a causal question; the results are the effects. The experiment is devised to test the causal relations; the results demonstrate the effect.

In an experimental situation or any other situation in which you investigate cause-effect relationships, you must mark out the boundaries of the investigated process. A cause can produce an effect which in turn becomes the cause of another effect and so on. The controlled conditions of an experiment attempt to delimit the boundaries of this process.

An example will elucidate the importance of defining the boundaries: One cause of a tidal wave is an earthquake at sea. The earthquake sends out seismic shocks which are distributed in the water, and since water moves more slowly than the sound waves, a mass of water builds up and strikes the land with a force that is greater than the force and fury of a normal wave

The effects of the tidal wave are many. Houses are destroyed, piers fall into the sea, boats are tossed and pulled from their moorings, effluents normally discharged through pipes are forced back up into pipes.

The cause of the tidal wave is the earthquake.

The effect of the tidal wave is the destruction of property.

The cause comes first. The effect is the outcome.

But, you might argue, the earthquake is *really* the cause of the destruction; the earthquake came first and it, ultimately, created the destruction. The destruction to property can be understood as the effect of a tidal wave in one scenario. However,in another scenario, the destruction could be construed as the final cause of the tidal wave.

# Kinds of Causes

### Proximate and Remote Causes

The previous example asserted that the tidal wave was the direct cause of property destruction. Another word for direct cause is *proximate*. The proximate cause in a chain of events is immediately prior to the event. What is the closest cause for the destruction? The tidal wave. While the earthquake causes the tidal wave, the earthquake ITSELF did not destroy the pier. The earthquake is a cause but not a direct cause; it is a *remote* cause because it is farther away from the situation than the proximate cause.

It is advisable to deal with proximate causes rather than remote ones to avoid many steps of backtracking. In order to deal with direct causes, you need therefore to carefully identify what change you want to investigate. If you deal with the tidal wave, it must be the center of your cause-effect analysis.

before — FOCUS — after

cause — situation — effect

earthquake — tidal wave — destruction

If you focused on the earthquake, you would come up with a completely different analysis. The cause of the earthquake is shifting tectonic plates in the earth; the effect of the earthquake is the tidal wave.

FOCUS

cause — situation — effect

earth tectonics — earthquake — tidal wave

Let us change this around one more time, and place the destruction of property in the central position. The cause of mass destruction is the tidal wave; the effect of this destruction is that businesses close, people lose money, schools are closed.

FOCUS
cause — situation — effect
tidal wave — destruction — businesses closed

When you write a process analysis, be very precise in selecting the term you emphasize. A slight shift of focus will entirely alter your essay.

Unless otherwise specified, deal with proximate causes rather than remote ones. When you deal with causes so remote that you engage ultimate causes or first causes (Why was I born? Why is there life on earth?), you are in a realm of philosophy known as metaphysics—*physics* is the Greek word for "nature"; *meta-* means "transcending or comprehending." According to *The American Heritage Dictionary*, metaphysics, based on speculative or abstract thought, "is a branch of philosophy which investigates the nature of first principles and problems of ultimate reality including the study of being (ontology) and, often, the study of the structure of the universe (cosmology)."

Some guidelines for proposals will make the following specification: "State the broad, long-term objectives (of your research project) and describe concisely and realistically what the specific research described in this application is intended to accomplish and any hypotheses to be tested." Specifications for the background ask you to "State concisely the importance of the research described in this application by relating the specific aims to the broad, long-term objective." In asking you to define the long-term objectives of your research work and how the present research will accomplish those aims, you are asked to situate the proposed research within a cause-effect relationship. In other words, what immediate effect will your research produce, and what long-range effect may be possible?

Long-range effects may difficult to assess because you have no way of knowing how your paper may contribute to the body of received knowledge in the field or may stimulate others working in a related field. Beyond a certain reasonable range of extension, such matters fall into weaker causal relations of influence than direct cause. Be circumspect in describing the long-range effects of your research.

### Necessary, Contributory, and Sufficient Causes

In addition to proximate and remote causes, there are three other kinds of causes: necessary, contributory, and sufficient. The categories are not mutually exclusive; it is possible to have a proximate necessary cause and a proximate sufficient cause.

A *necessary* cause is one that must be present for an effect to occur but by itself cannot cause the occurrence of the effect. Water is a necessary cause of a tidal wave; you have to have water in order to have a tidal wave. But other factors have to be involved before a normal wave turns into a tidal wave.

A *contributory* cause may produce an effect but cannot produce that effect alone. While good training may *contribute* to the success of an athlete, good training by itself cannot take the place of good reflexes, a strong constitution, and a knowledge of the game. A storm may contribute to the effect of a tidal wave but cannot be a greater force than that produced by a storm is necessary to create a tidal wave.

A *sufficient* cause can produce the effect by itself. A heart attack alone can kill even though a person has an ulcer, a sprained ankle, and a sore throat. A heart attack is a sufficient cause of mortality.

Most causes are not sufficient; most are a combination of necessary and contributory. When you explain causes which are not sufficient, you must explain how the causes interrelate.

It is appropriate at this point to offer another ordering of causes, derived from a classical Greek philosopher, Aristotle.

Aristotle formulated four types of causes: efficient, material, formal, and final.

The *efficient* cause of an event or thing is that which stops or starts it. The efficient cause is explicated by specifying its place in time. For example, the earthquake is the efficient cause of the tidal wave. The matter of a thing is its *material* cause: The material cause of the tidal wave is water. The *formal* cause ascribes causal relations to essence or nature. The behavior of water under conditions of stress is the formal cause of the tidal wave. The *final* cause is the ultimate end for which is a thing is made or a process started. The final cause of the earthquake is the release of stresses in the earth.

In a Darwinian view of evolution, the final cause is the continuation of a species. Another word for end (understood as a final point) is *telos*; explanations which concern themselves with the final goal are called teleological. Teleological explanations are framed in different ways; from the standpoint of evolutionary biology, teleology relates to survival of the species. In other disciplines like philosophy or metaphysics, teleologic concerns can be framed so broadly as to include: "What is the final meaning of life?"

Explanations about the final cause may be considered teleological; speculation about the ultimate end of processes or things may take one into the realm of philosophy rather than science. While the final cause of the acorn may be said to be the oak, this question could be complicated by asking what is the final cause of the oak?

In discussions about final causes, it is tempting to impute agency to an inanimate object, as in the acorn *desires, aspires, wants to be, wills itself* to be an oak. This inferred agency underlies theories of the universe which impute celestial and terrestrial motion to a Prime Mover; a universe conceived as a Grand Design implies a Grand Designer. While not to engage too fully with metaphysical first causes, it should be pointed out that English, and romance language grammar, by virtue of the nature of the language, always permits the creation of a linguistic agent. Nearly every verb can be converted into a noun of agency: do/doer, go/goer, think/thinker, design/designer, pattern/patterner. Our language, save in rare cases, does not permit us to think of an action without a doer of the action; an exception to this is the word happen. We do not speak of a "happener."

# Effects

### Analyzing Effects Using Comparison/Contrast

*1. Identification of the Effect.*
   Precisely identify the effect you want to examine.
   Your focus is extremely important. Distinguish between effects and side-effects, between products and by-products, between the direct effect of a cause and the effect of an effect.
   When the orthopedic surgeon Paul Brand went to Vellore, India, in 1947, he wanted to use reconstructive surgery to treat the ailments of lepers. Among the effects of leprosy were blindness, clawed limbs, loss of fingers and toes. While the *bacillus leprae* directly produced tubercles which destroyed the delicate nerve endings responsible for sensation, there was no reason to account for the sloughing-off of body parts. After he shook hands with a leper, Dr. Brand noticed that the leper possessed unusual strength; he came to find that the strength itself was not unusual but that, in the absence of sensory input, the leper could feel neither pressure nor pain. In fact, limbs were lost in the course of everyday errands because the leper could not feel pain. Some lepers lost extremities to the rats which infested the buildings at night. Loss of feeling led to loss of limbs. Furthermore, lepers acquired eye infections and ulcers because, in the absence of sensation, the lepers did not blink away irritating dust or foreign bodies. Blinking is also an eye-washing reflex; without this response, the lepers suffered eye infections. While the direct effect of the bacillus was the production of tubercles, blindness and limb loss were the effects of effects.

*2. Discuss the end with reference to the beginning.*
   Start with the end of the process first. The results or the process are the terminal point of the process; you want to explain how you got there. In order to explain how you arrived at a point, you need to explain where you started. For instance, if I told you my airline flight ended at Los Angeles, you would not have very much information about types of airlines, routes, or itineraries, length or significance of the journey unless you knew where I started. A trip that starts at New York City and a trip that starts at Mexico City are completely different trips. You discuss the end with respect to the beginning in order to pinpoint the change.
   To describe a change, look at what happened before the change, what

happened after the change, and explain the events or phenomena that intervened to make the end result from the beginning. In scientific experimentation, this takes the form of comparing your hypothesis with your experimental results — how are they alike? how are they different?

## False Cause

When engaging in a cause-effect analysis, be careful to avoid errors in reasoning called fallacies. The fallacy apposite to this discussion is called false cause. This fallacy describes an error in attributing cause to something which had no causal relation. To ascribe a causal relation to things which are correlated is fallacious. For instance, while there may be a correlation between men who eat Cheerios and men who do not get pregnant, there is no causal relationship to suggest that Cheerios prevent men from getting pregnant. A related fallacy is called *post hoc, ergo propter hoc*. This fallacy refers to ascribing causal relations to events related in time. The passenger who said, "I should have left the lights on," after he turned off the stateroom lamp and the Titanic sank is making such an attribution. This passenger asserted that because he turned off the lights, the Titanic sank.

## Exercises

### 1. Changing Orientation

The following passage comes from a sixteenth-century book on magnetism. Although you may find the language a bit strange, the following passage contains a very orderly set of instructions. Rewrite Gilbert's essay into a set of instructions in modern English with headings. Make sure that you have an introduction and a conclusion; sufficient information in this passage permits you to write both.

William Gilbert was an English doctor born in England in 1540 (d. 1603). His contemporaries were Shakespeare (1564–1616), Galileo Galilei (1564-1642), and Andreas Vesalius (1514–1564). In 1573, Gilbert was elected to the Royal College of Physicians. He became physician-in-ordinary to Queen Elizabeth I, who so favored him that she settled an annual pension on him (said to be the only legacy left by her to anyone).

During his studies, Gilbert was first interested in chemistry but later turned to magnetism and electricity, which, while known in the east, had received no discussion in western civilization since the time of Thales. Gilbert's book *De magnete* caused no stir in England but was received enthusiastically in Europe, where it is said Galileo turned to studying magnetism after reading it. Although Dr. Priestley called Gilbert "the father of modern electricity," Gilbert's work is more important for its emphasis on the inductive method based on the observation of physical phenomena. This method was a relatively new idea at the time; the previous method of science had been based on the authority of the ancients.

## Chapter IV. Which Pole Is the North:
## How the North Pole Is Distinguished from the South Pole

One of the earth's poles is turned toward Cynosura [constellation] and steadily regards a fixed point in the heavens (save that it is unmoved by the precession of the fixed stars in longitude, which movement we recognize in the earth, as we shall later show); the other pole is turned toward the opposite aspect of the heavens, an aspect unknown to the ancients, but which is adorned with a multitude of stars, and is itself a striking spectacle for those who make long voyages. So, too, the loadstone possesses the virtue and the power of directing itself toward the north and the south (the earth itself co-operating and giving to it that power) according to the conformation of nature, which adjusts movements of the stone to its true locations. In this manner it is demonstrated: Put the magnetic stone (after you have found the poles) in a round wooden vessel—a bowl or dish; then put the vessel holding the magnet (like a boat with a sailor in it) in a tub of water or cistern where it may float freely without touching the rim, and where the air is not stirred by winds (currents) which might interfere with the natural movement of the stone: there the stone, as if in a boat floating in the middle of an unruffled surface of still water, will straightaway set itself, and the vessel containing it in motion, and will turn in a circle till its south pole shall face north and its north pole, south. For, from a contrary position, it returns to the poles; and though with its first too strong impetus it passes beyond, still as it comes back again and again, at last it rests at the poles or in the meridian (save that, according to the place, it diverges a very little from those points, for from the meridional line, the cause of which we will define later). As often as you move it out of its place, so often, by reason of the extraordinary power with which nature has endowed it, does it seek again its fixed and determinate points. Nor does this only occur only when the poles in the loadstone are made to lie evenly in the plane of the horizon; it takes place also even though one pole, whether north or south, be raised or depressed 10, 20, 30, 40, or 80 degrees from the plane of the horizon; you shall see the north part of the stone seek the south, and the south part the north. . . . Further, it is to be remembered that all who have hitherto written about the poles of the loadstone, all instrument-makers and navigators, are egregiously mistaken in taking for the north pole of the loadstone the part of the stone that inclines to the north, and for the south pole the part that looks to the south; this we will hereafter prove to be an error. So ill-cultivated is the whole philosophy of the magnet still, even as regards its elementary principles.[1]

## 2. Correcting Instructions: Blowing Up Caterpillars

The following are a set a directions for preparing an insect specimen, a caterpillar, for a collection box. But these directions need revision at the level of content and at the level of grammar and punctuation. Fill in any missing parts. Check the equipment list against the procedure—is it complete and ordered by use? Use headings to divide the text and use headings to separate the steps. Are all warnings or notes put in the appropriate place? Are there are any steps which should have warnings?

### Blowing Up Caterpillars

Caterpillars are juicy, and if left to dry out by themselves, they will rot, discolor, and smell. To preserve them one must make sure that all of the organisms of decomposition (bacteria) are excluded because the environment is made unsuitable for them to grow and reproduce. The caterpillar may be placed in a chemical preservative such as 70 isopropyl alcohol or a solution of formaldehyde, or the flesh of the caterpillar that may support bacterial life can be removed and the remaining exoskeleton dried.

In this exercise, you will learn how to remove the juicy caterpillar contents and dry the remaining exoskeleton in a somewhat lifelike appearance.

*Equipment*: drying oven; alcohol lamp; matches; canula; wooden dowel; caterpillar pump and hose; balsa wood scraps; insect pin; fingernail polish (clear).

*Procedure*: With fine scissors cut around the anal opening of freshly killed caterpillar. If the larva has poisonous spines or hairs handle with caution. Lay the caterpillar on paper towel and gently and carefully roll the wooden dowel over the caterpillar from behind the head toward the anus. This process will squeeze the contents of the caterpillar out of the opening that you cut around the anus. This should empty the "bag" (the caterpillar) without breaking the skin. Make sure that you point the anal end of the caterpillar away from you. Now insert the tapered point of the canula into the anal opening of the caterpillar and hold the larva in position with the spring clips. Attach the end of the rubber hose of the pump to the canula and pump the caterpillar with air until wrinkles and bands are smoothed. Make any minor adjustments that are necessary with forceps but do not puncture the larvae. Insert the caterpillar now on the end of the canula into the oven and observe it through the glass plate that forms the top of the oven. Turn the caterpillar in spit-like fashion until it is dry. As soon as it holds its shape upon release of the pressure in it you can remove it from the canula.

Insert a piece of balsam into the anal opening and glue it there with the fingernail polish, Duco cement or shellac. Put the insect pin in the balsam plug and attach labels to the pin. The caterpillar is now ready for the collection drawer.[2]

## Notes

1. William Gilbert, *De Magnete*, trans. P. Fleury Mottelay (New York: Dover Publications, 1958); this is an unaltered republication of the 1893 translation), pp. 26–27.
2. This exercise was developed by Professor Wayne Losano, University of Florida.

# Reading Selections

# Take This Fish and Look at It

## Samuel H. Scudder

Samuel H. Scudder (1837–1911) was an American scientist who was educated at Williams College and Harvard University. One of the most learned entomologists of the day, his main scientific contributions were in the study of butterflies and Orthoptera (an order of insects that includes grasshoppers and crickets). Scudder was a student of Louis Agassiz (1807–1873), the distinguished Harvard professor of natural history, who used to subject his students to a rigorous but useful exercise in minute observation.[1] The following is Scudder's account of one such exercise.

It was more than fifteen years ago that I entered the laboratory of Professor Agassiz, and told him I had enrolled my name in the Scientific School as a student of natural history. He asked me a few questions about my object in coming, my antecedents generally, the mode in which I afterwards proposed to use the knowledge I might acquire, and finally, whether I wished to study any special branch. To the latter I replied that, while I wished to be well grounded in all departments of zoology, I purposed to devote myself specially to insects.

"When do you wish to begin?" he asked.

"Now," I replied.

This seemed to please him, and with an energetic "Very well!" he reached from a shelf a huge jar of specimens in yellow alcohol. "Take this fish," he said, "and look at it; we call it a haemulon; by and by I will ask what you have seen."

With that he left me, but in a moment returned with explicit instructions as to the care of the object entrusted to me. "No man is fit to be a naturalist," said he, "who does not know how to take care of specimens."

I was to keep that fish before me in a tin tray, and occasionally moisten the surface with alcohol from the jar, always taking care to replace the stopper tightly. Those were the days of ground-glass stoppers and elegantly shaped exhibition jars; all the old students will recall the huge neckless glass bottles with their leaky, wax-besmeared corks, half eaten by insects, and begrimed with cellar dust. Entomology was a cleaner science than ichthyology, but the example of the Professor, who had unhesitatingly plunged to the bottom of the jar to produce the fish, was infectious; and though this alcohol had a "very ancient and fishlike smell," I really dared not show any aversion within these sacred precincts, and treated the alcohol as though it were pure water. Still, I was conscious of a passing feeling of disappointment, for gazing at the fish did not commend itself to an ardent entomologist. My friends at home, too, were annoyed when they discovered that no amount of eau-de-cologne would drown the perfume which haunted me like a shadow.

In ten minutes I had seen all that could be seen in that fish, and started in search of the Professor—who had, however, left the Museum; and when I returned, after lingering over some of the odd animals stored in the upper apartment, my specimen was dry all over. I dashed some fluid over the fish as if to resuscitate the beast from a fainting fit, and looked with anxiety for a return of the normal sloppy appearance. This little excitement over, nothing was to be done but to return to a steadfast gaze at my mute companion. Half an hour passed—an hour—another hour; the fish began to look loathsome. I turned it over and around; looked it in the face—ghastly; from behind, beneath, above, sideways, at a three-quarters' view—Just as ghastly. I was in despair; at an early hour I concluded that lunch was necessary; so, with infinite relief, the fish was carefully replaced in the jar and for an hour I was free.

On my return, I learned that Professor Agassiz had been at the Museum, but had gone, and would not return for several hours. My fellow-students were too busy to be disturbed by continued conversation. Slowly I drew forth that hideous fish, and with a feeling of desperation again looked at it. I might not use a magnifying-glass; instruments of all kinds were interdicted. My two hands, my two eyes, and the fish; it seemed a most limited field. I pushed my finger down its throat to feel how sharp the teeth were. I began to count the scales in different rows, until I was convinced that was nonsense. At last a happy thought struck me—I would draw the fish; and now with surprise I began to discover new features in the creature. Just then the Professor returned.

"That is right," said he; "a pencil is one of the best of eyes. I am glad to notice, too, that you keep your specimen wet, and your bottle corked."

With these encouraging words, he added, "Well, what is it like?"

He listened attentively to my brief rehearsal of the structure of parts whose names were still unknown to me; the fringed gill arches and movable operculum; the pores of the head, fleshy lips and lidless eyes; the lateral line, the spinous fins, and forked tail; the compressed and arched body. When I finished, he waited as if expecting more, and then with an air of disappointment:

"You have not looked very carefully; why," he continued more earnestly, "you haven't even seen one of the most conspicuous features of the animal, which is as plainly before your eyes as the fish itself; look again, look again!" and he left me to my misery.

I was piqued; I was mortified. Still more of that wretched fish! But now I set myself to my task with a will, and discovered one new thing after another until I saw how just the Professor's criticism had been. The afternoon passed quickly, and when, towards its close, the Professor inquired: "Do you see it yet?"

"No," I replied, "I am certain I do not, but I see how little I saw before."

"That is next best," said he, earnestly, "But I won't hear you now; put away your fish and go home; and perhaps you will be ready with a better answer in the morning. I will examine you before you look at the fish."

This was disconcerting. Not only must I think of my fish all night, studying, without the object before me, what this unknown but most visible feature might be; but also, without reviewing my discoveries, I must give an exact account of them the next day. I had a bad memory; so I walked home by the Charles River in a distracted state, with my two perplexities.

The cordial greeting from the Professor the next morning was reassuring; here was a man who seemed quite as anxious as I that I should see myself what he saw.

"Do you perhaps mean," I asked, "that the fish has symmetrical sides with paired organs?"

His thoroughly pleased, "Of course! Of course!" repaid the wakeful hours of the previous night. After he had discoursed most happily and enthusiastically—as he always did—upon the importance of this point, I ventured to ask what I should do next.

"Oh, look at your fish!" he said, and left me again to my own devices. In a little more than an hour, he returned and heard my new catalogue.

"That is good, that is good!" he repeated; "but that is not all; go on"; and so for three long days he placed that fish before my eyes, forbidding me to look at anything else, or to use any artificial aid. "Look, look, look," was his repeated injunction.

This was the best entomological lesson I ever had—a lesson whose influence has extended to the details of every subsequent study; a legacy the Professor had left to me, as he has left to so many others, of inestimable value, which we could not buy, with which we cannot part.

A year afterward, some of us were amusing ourselves with chalking outlandish beasts on the Museum blackboard. We drew prancing starfishes; frogs in mortal combat; hydra-headed worms; stately crawfishes standing on their tails, bearing aloft umbrellas; and grotesque fishes with gaping mouths and staring eyes. The Professor came in shortly after, and was amused as any at our experiments. He looked at the fishes.

"Haemulons, every one of them," he said; "Mr. ____drew them."

True, and to this day, if I attempt a fish, I can draw nothing but haemulons.

The fourth day, a second fish of the same group was placed beside the first, and I was bidden to point out the resemblances and differences between the two; another and another followed, until the entire family lay before me, and a whole legion of jars covered the table and surrounding shelves; the odor had become a pleasant perfume; and even now, the sight of an old, six-inch worm-eaten cork brings fragrant memories.

The whole group of haemulons was thus brought in review; and, whether engaged upon the dissection of the internal organs, the preparation and examination of the bony framework, or the description of the various parts, Agassiz's training in the method of observing facts and their orderly arrangement was ever accompanied by the urgent exhortation not to be content with them.

"Facts are stupid things," he would say, "until brought into connection with some general law."

At the end of eight months, it was almost with reluctance that I left these friends and turned to insects; but what I had gained by this outside experience has been of greater value than years of later investigations in my favorite groups.

## Writing Exercises

### 1. Observation

Take an object, a plant, an animal, or a seashell and follow the same procedure as Scudder.

### 2. Statement of Purpose

When Agassiz asked Scudder what he intended to do with his knowledge, Scudder said he proposed to be an entomologist. Write a statement of purpose in which you, adopting Scudder's persona, explain why you wish to study with Professor Agassiz and what you can learn from him.

### 3. Analysis

Outline the steps in Scudder's increasingly focused observations. Is there a pattern to these steps? Does Scudder proceed inductively or deductively?

### 4. Explanation

(a) Agassiz says: "A pencil is one of the best of eyes." Explain. Can you interpret this in light of the difference between casual observation and focused observation described in Chapter 1?

(b) What does Agassiz mean by the point that "Facts are stupid things until brought into connection with some general law"?

## Notes

1. Samuel L. Scudder, "Take This Fish and Look at It," in Jo Ray McCuen and Anthony C. Winkler, *Reading for Writers*, 4th ed. (New York, San Diego, Chicago: Harcourt, Brace, and Jovanovich, Inc., 1983, 1980, 1977, 1974), pp. 82–85.

# Epidemics, Book I

## Hippocrates

Hippocrates* (c. 430 B.C.) was a famous Greek physician who is considered the father of medicine. Even today, practicing physicians take the Hippocratic oath, a code of medical ethics, which is based on the precept: *First, do no harm.* Hippocrates wrote extensively on many subjects, including epidemics, embryology, and the influence of geographical location on disease. He argued that epilepsy, the "sacred disease," was not caused by possession of the gods. He prescribed practical treatment for fractures and wrote an anatomy of the heart which was important for its careful and direct observation. Even though his writings were not addressed to a specialized audience of medical students but to the general public of his time, his influence among men of science extended through the sixteenth, seventeenth, and eighteenth centuries.

---

\* Hippocrates, writing 400 years before the introduction of the Julian calendar, had no convenient method for recording dates. There were several different calendars in use in Greece, and they were all based on lunar months, so that the same date would not always fall on the same day of the solar year. It was therefore common practice to use certain astronomical events as a rough method for dating. The four obvious points are the equinoxes and solstices: 21 March, 21 September, 21 June, 22 December. These are from time to time supplemented by reference to a heliacal rising or setting of certain stars and constellations. Owing to the precession of the equinoxes these are not constant, and various factors prevent an exact calculation of the dates Hippocrates intended. The chief of these mentioned in the text, together with their approximate equivalents are:

The rising of Arcturus . . . . . . 10 September
The rising of the Pleiads . . . . 10 May
The setting of the Pleiads. . . . 11 November
The rising of the Dog Star . . . 17 July

(J. Chadwich and W. N. Mann, translators, *Hippocratic Writings*, edited by G. E. R. Lloyd [London: Penguin Books, 1983], pp. 65–66.)

Keep in mind that in Hippocrates' time, any person could claim to be a physician. There were no medical schools or state licensing. In fact, there was, in those days, no category of "science" at all; the study of philosophy included ethics, logic and nature. Because the study of medicine was considered part of the study of nature, it was considered an appropriate subject for philosophers rather than for doctors. Even the philosopher, Plato, wrote theories on the origin of disease in his work the *Timeaus*. —G. E. R. Lloyd, editor's introduction to *Hippocratic Writings*.

Hippocrates wrote three books on epidemics; this excerpt is taken from Book 1 and deals with the fever he calls *causus*.

## Epidemics, Book I
### (iii)

13. A little before the rising of Arcturus and during its season there were many violent rainstorms in Thasos accompanied by northerly winds. About the time of the equinox and until the settling of the Pleiads, the winds were southerly, so little rain fell. The winter was northerly with periods of drought, cold, high winds, and snow. There was very severe storms at the time of the equinox. The spring was northerly, dry, with a little rain, and it was cold. There was a little rain at the time of the summer solstice but instead a severe cold spell set in and lasted till the [rising of the] Dog Star. Thence, until the [rising of] Arcturus, the summer was hot. This hot spell began suddenly and was both continuous and severe. There was no rain and the etesian winds began and continued until the equinox.

14. Under such circumstances, cases of paralysis started to appear during the winter and became common, constituting an epidemic. Some cases were swiftly fatal. In other respects, health remained good. Cases of *causus* were encountered early in the spring and continued past the equinox towards the summer. Most of those who fell sick in the spring or at the very beginning of summer recovered, though a few died. In the autumn, when the rains came, the disease was more fatal and the majority of those that took it died.

It was a peculiarity of *causus* that a good copious epistaxis often proved a cure, and I do not know of any in these circumstances who died if they had a good epistaxis. For Philiscus, Epameinon and Silenus had a small epistaxis on the fourth and fifth days; they died. Most of those who were sick had shivering attacks about the time of the crisis, especially those who did not have epistaxis. Such patients also had attacks of sweating.

15. Some cases of *causus* developed jaundice on the sixth day and these were assisted by the evacuation of urine, abdominal disturbance or by a profuse hemorrhage, such as Heracleides (who lay at Aristocydes' house) had. Moreover in this case he did not only have epistaxis but trouble in the belly and diuresis as well. He reached a crisis on the twentieth day. The

servant of Phanagras was not so lucky; he had none of these things happen to him and he died.

Most patients suffered from hemorrhage and especially was this the case in youths and young men. Indeed, of the latter who did not have a hemorrhage, most died. In older people the disease turned to jaundice or their bellies were upset, as was the case of Bion, who lay at Silenus' house. During the summer dysentery became epidemic and those who had not recovered by that time had their sickness end up as a sort of dysentery, even when they had a hemorrhage. This happened to Myllus and to Erato's slave, whose illness, after a copious hemorrhage, turned to a sort of dysentery; they survived.

In fact, in this disease, this fluid was peculiarly abundant. Even those who did not bleed about the time of the crisis suffered pain and passed thin urine at this time and then began to bleed slightly about the twenty-fourth day, and there was pus mixed with blood. In the case of Antiphon the son of Critobulus this finally ceased and the ultimate crisis was reached about the fortieth day. Such cases showed hard swellings near the ears which absorbed and were followed by a heaviness in the left flank and in the region of the iliac crest.

16. Many women were sick, but fewer women than men, and the disease in them was less fatal. Childbirth was often difficult and was followed by disease. These cases were specially fatal as, for instance, in that of the daughter of Telebulus, who died on the sixth day after giving birth. In most cases, bleeding from the womb occurred during the fever and in many girls it occurred for the first time, but some had epistaxis. In some cases both bleeding from the womb and epistaxis were observed. For instance, the daughter of Daitharses who was a virgin not only had uterine bleeding for the first time then but also had a violent discharge of blood from the nose. I know of no case which proved fatal if either of these complications ensued. So far as I know, all who fell ill while pregnant aborted.

17. Generally, in this disease, the urine was of good color but thin with a slight sediment. The belly was disordered, the stools being thin and bilious. In many cases, after a crisis had been reached for other disorders, the malady ended up as dysentery, as happened to Xenophanes and Critias. I will record the names of those patients who had watery, copious and fine urine, even after a crisis, with a healthy sediment, and who had a favourable crisis in other respects too. They were Bion, who lay at the home of Silenus; Cratis, who was at the Xenophanes; the slave of Areto, and the wife of Mnesistratus. All these subsequently suffered from dysentery.

About the time of Arcturus many reached the crisis on the eleventh day and they did not suffer the expected relapses. About this time, especially in children, the malady was associated with coma and these cases were the most rarely fatal of all.

18. *Causus* lasted on to the equinox, up to the setting of the Pleiads, and

even into the winter. But at this time brain fever became prevalent and most of its victims died. A few similar cases were also seen during the summer. Those suffering from fever of the *causus* type which proved fatal showed certain additional symptoms even at the beginning of the illness. High fever attended the beginning of the illness along with slight shivering fits, insomnia, thirst, nausea, and a little sweating about the forehead and over the clavicles (in no cases all over), much delirium, fears, and despondency, while the extremities such as the toes were chilled, but especially the hands. Paroxysms occurred on even days. Generally, pain was greatest on the fourth day and the sweat was cold. Their extremities did not regain warmth but remained cold and livid, and they no longer suffered from thirst. They passed little urine, which was black and fine, and became constipated. In none of these cases was there a discharge of blood from the nose but only a few drops. Nor did these cases show any remission but died on the sixth day, sweating. Those patients who developed brain fever had all the above symptoms, but the crisis usually took place on the eleventh day. Where brain fever was not present at the beginning but appeared on the third or fourth day, the crisis did not take place until the twentieth day. In these the illness was moderate in its severity first but became severe about the seventh day.

19. The disease was very widespread. Of those who contracted it, death was most common among youths, young men in the prime of life, among those with smooth skins, those of a pallid complexion, those with straight hair, those with black hair, those with black eyes, those who had been given to violent and loose living, those with thin voices, those with rough voices, those with lisps and choleric. Many women also succumbed to this malady. During this epidemic there were four signs which betokened recovery: a considerable epistaxis, a copious discharge of urine that contained a lot of favorable sediment, biliousness and disorders of the belly coming on a favourable crisis, or if there were dysentery. . . .

If a patient weeps in spite of himself in acute fever of the type of *causus*, you must expect an epistaxis, even if there is no reason to expect a fatal outcome. If a patient be poorly, it portends not hemorrhage but death. . . .

23. The factors which enable us to distinguish between diseases are as follows: First we must consider the nature of man in general and of each individual and the characteristics of each disease. Then we must consider the patient, what food is given him and who gives it—for this may make it easier for him to take or more difficult—the conditions of the climate and locality both in general and in particular, the patient's customs, mode of life, pursuits and age. Then we must consider his speech, his mannerisms, his silences, his thoughts, his habits of sleep or wakefulness and his dreams, their nature and time. Next, we must note whether he plucks his hair, scratches or weeps. We must observe his paroxysms, his stools, urine, sputum, and vomit. We look for any changes in the state of his malady, how often such changes occur and their nature, and the particular changes which

induce death or a crisis. Observe too, sweating, shivering, chill, cough, sneezing, hiccough, the kind of breathing, belching, wind, whether silent or noisy, hemorrhages and hemorrhoids. We must determine the significance of all these signs.

## Exercises

### 1. Stipulative Definition

In the previous chapter on definition, you learned how to write three kinds of definition. These definitions were used to offer an observation about a thing or event in the world; those definitions were used for items which had a determinant reference in the world. But sometimes you will find that you encounter a phenomenon for which you have no referent, although the person writing the description or the report had a clear idea of what he or she saw. In such cases, you cannot construct a definition based on what you know; you must proceed in a different manner by considering what the word means in the context of the writing. When you define a word by the meanings which accrue to it through context, you write a **Stipulative Definition**. A stipulative definition explains the way that meaning is stipulated through the writing. While you may think you know what the person describes, you may not be entirely sure. Write a definition on the basis of text rather than on the basis of reference to a known event or object in the world.

*Causus* is such a term. Some people think that Hippocrates was blending descriptions of different illnesses when he used the word *causus*. Read the passage from Hippocrates again and write an extended stipulative definition for *causus*. Organize your description using the techniques you have already learned and which are listed in the Definition section. Consider the cause, the spread, and the symptoms of the disease. Consider also what Hippocrates means by the crisis.

### 2. Comparison/Contrast

Hippocrates claims that it is important to consider aspects of the person in the pathology of the disease. He notes that the disease leads most commonly to death among those with thin voices, those with rough voices, etc. Look at all the passages in which Hippocrates deals with the personal life of the patient, not only in his description of the pathology but also in his list of those who died.

Analyze the role of the individual in the pathology of disease according to Hippocrates; then read Fracastoro and do the same for him. Finally, write an essay in which you compare/contrast the significance of the individual person in the spread of disease. Would Thucydides' account of the plague of Athens be closer to Hippocrates or to Fracastoro?

### 3. Inference

Read Hippocrates' account carefully and then write an essay in which you explain what model of diseases underlies his description. Unlike Fracastoro, he does not rely upon the germ model of disease. How would you characterize Hippocrates' assumptions about what causes disease? What is the connection between weather and disease?

# The Plague of Athens

## Thucydides
## (c. 460–400 B.C.)

The following excerpt, "The Plague of Athens" (430 B.C.), is taken from Thucydides' history *The Peloponnesian War*. Thucydides was a general on the Athenian side against Sparta. He fell ill of the plague but later recovered. Thucydides was a careful observer, and even though he would not have possessed knowledge of the germ theory of disease (diseases are spread by infectious organisms), his description has much merit. —Thucydides, *History of the Peloponnesian War*, trans. Rex Warner (New York: Viking Penguin Books, 1985), pp. 152–55.

The plague originated, so they say, in Ethiopia in upper Egypt, and spread from there into Egypt itself and Libya and much of the territory of the King of Persia. In the city of Athens, it appeared suddenly, and the first cases were among the population of Piraeus. . . . As to the question of how it could have first come about or what causes can be found adequate to explain its powerful effect on nature, I must leave that to be considered by other writers, with or without medical experience. I myself shall merely describe what it was like, and set down the symptoms, knowledge of which will enable it again to be recognized, if it should ever break out again. I had the disease myself and saw others suffering from it.

That year, as it is generally admitted, was particularly free from all other kinds of illness, though those who did have any illness previously caught the plague in the end. In other cases, however, there seemed to be no reason for the attacks. People in perfect health suddenly began to have burning feelings in the head; their eyes became red and inflamed; inside their mouths there was bleeding from the throat and tongue, and the breath became unnatural and unpleasant. The next symptoms were sneezing and hoarseness of voice, and before long, the pain settled on the chest and was accompanied by coughing. Next the stomach was affected with stomach-aches and vomitings of every kind of bile that has been given a name by the medical profession, all this being accompanied by great pain and difficulty. In most cases there were attacks of ineffectual retching, producing violent spasms; this sometimes ended with this stage of the disease, but sometimes continued long afterward. Externally, the body was not very hot to the touch, nor was there any pallor: the skin was rather reddish and livid, breaking out in small pustules and ulcers. But inside there was a feeling of burning, so that people could not bear the touch even of the lightest linen clothing, but wanted to be completely naked, and most of all would have liked to plunge into cold water. Many of the sick who were uncared for actually did so, plunging into the water-tanks [cisterns] in an effort to relieve a thirst which was unquenchable; for it was just the same with them whether they drank much or little. Then all the time they were afflicted with insomnia and the desperate feeling of not being able to keep still.

In the period when the disease was at its height, the body, so far from wasting away, showed surprising powers of resistance to all the agony, so that there was still some strength left on the seventh or eighth day, which was the time when, in most cases, death came from the internal fever. But if people survived this critical period, then the disease descended to the bowels, producing violent ulcerations and uncontrollable diarrhea, so that most of them died later as a result of the weakness caused by this. For the disease, first settling in the head, went to affect every part of the body in turn, and when people escaped its worst effects, it still left its traces on them by fastening upon the extremities of the body. It affected the genitals, the fingers, the toes, and many of those who recovered lost the use of those members. There were some also who, when they first began to get better, suffered from a total loss of memory, not knowing who they were them-selves and being unable to recognize their friends. . . .

Terrible was the sight of people dying like sheep through having caught the disease as a result of nursing others. This indeed caused more deaths than anything else. . . . Yet the ones who felt most pity for the sick and the dying were those who had the plague themselves and recovered from it. They knew what it was like and at the same time felt themselves to be safe, for no one ever caught the disease twice, or if he did, the second attack was never fatal.

A factor which made matters worse than they already were was the removal of people from the country into the city, and this particularly affected the in-comers. There were no houses for them, and living as they did during the hot season in badly ventilated huts, they died like flies. The bodies of the dying were heaped on top of the other, and half-dead creatures could be seen staggering about in the streets or flocking around the foun-tains in their desire for water. The temples . . . were full of the dead bodies of people who had died inside them. For the catastrophe was so overwhelming that men, not knowing what would happen next to them, became indifferent to every rule of religion or law. All the funeral ceremonies which used to be observed were now disorganized, and they buried the dead as best as they could. Many people, lacking the necessary means of burial because so many deaths had already occurred in their households, adopted the most shameless methods. They would arrive first at a funeral pyre that had been made by others, put their own dead on it, and set it alight; or, finding another pyre burning, they would throw the corpse that they were carrying on top of the other and go away.

## Reading Questions

a. How does Thucydides explain his reason for describing the plague?

b. Outline the course of the symptoms as they are described in the second paragraph. How does Thucydides organize his description? (By time, by space, from inside to outside; from outside to inside; from beginning to end?)

c. Although Thucydides' greater pattern of organization is arranged temporally—how the disease appears in time, from its onset in the patient to the patient's death—there are other patterns within this larger temporal account. Identify some of these patterns.

d. Consider what is missing from this description: What parts are left out? Why does he continue his description beyond the death of the patient? Why does he not describe an autopsy or a pathology report? Why does he not describe microbes or bloodwork?

e. What does Thucydides think caused the disease?

f. In most scientific reports, we expect the writer to remain objective, but Thucydides occasionally says things that indicate his personal view. Identify passages where Thucydides seems to make a subjective remark.

g. Thucydides suggests that the effects of the plague were intensified by the fact that it occurred within the city of Athens. How are the effects of the plague concentrated in the city? How do population dynamics affect its transmission?

h. What are the sociological effects of the disease? What happens to society as a consequence of the disease? What rituals or celebrations change? Let us suppose that you, like Thucydides or other Greeks, had no knowledge of the progress of the disease. How would you interpret this disaster in the city?

## Exercises

### 1. Definition
You think that the plague which Thucydides describes might be either typhoid

or typhus. When you look up the definition for these diseases, you find the following:

> **Typhoid fever:** A highly infectious, septicemic disease of humans caused by *Salmonella typhi* which enters the body by the oral route through ingestion of food or water contaminated by contact with fecal matter.
>
> **Typhus fever:** Any of three louse-born human diseases caused by *Rickettsia prowazakii* characterized by fever, stupor, headaches, and a dark-red rash.

Compare (contrast) these definitions with the passage. Do the definitions give sufficiently limiting criteria for you to identify the disease as one or the other? If these definitions are insufficient, look up the disease and complete the definition.

### 2. Formal Report

By this point, either in your notes or in your mind, you have compared and contrasted the characteristics of typhoid and typhus with the description of the plague in Athens.

Meanwhile, the Center for Disease Control has expressed interest in your diagnosis: The Center has requested that you send them a report of your comparison/contrast. Although the CDC will eventually want to see your entire report, they specifically asked you to fax them the "Discussion" section of your paper.

Write a "Discussion" section for a formal paper in which you identify the disease and explain your reasoning.

# The Germ Theory of Disease

## Girolamo Fracastoro
### (1546)

An infectious (or contagious) disease is a *process* (not a thing) that occurs in a host as a result of an interaction with a parasite. The parasite isolated from the host is not the disease but merely the potential causer of the disease. The disease itself is a complex interaction between parasite and host. With a given host and a given parasite, under the usual environmental conditions, a given disease usually arises. This disease is recognized in the host because of certain symptoms which become evident to an observer. In many, if not most cases, two hosts of the same species usually exhibit similar symptoms when infected with a given parasite. Because of this, we can recognize the same disease in several or many individual hosts.

It was only these symptoms which were available to early observers. They had no knowledge of microorganisms, and even little knowledge of the nature of the host. But because a number of diseases, such as syphilis, plague, tuberculosis, small pox, usually had characteristic symptoms, it was possible to observe these diseases in populations of individuals. By such observation, it was possible to infer these diseases were transmitted from person to another. Remember that the agency of this transfer was unknown. It was merely known that a transfer occurred. . . .

Fracastoro's writing are mainly philosophical and he attempts to define the subject in terms of the ideas of the day. Even though his discussions seems crude by current standards . . . [they come] close . . . to hitting the nail on the head. . . .

This paper was written in Latin. It is interesting that Fracastoro uses a word for the infectious or contagious principle which can best be translated in English as "germ.". . . His statement that these germs generate and propagate other germs is prophetical although not based on direct observation. Finally, his attempt to find a similarity between putrefaction and contagion is noteworthy because it was this very similarity which led later workers, such as Henle, Lister, Pasteur, and Koch, to consider seriously the germ theory of disease. —Thomas Brock, editor and translator *Milestones in Microbiology* (Washington, D.C.: American Society for Microbiology, 1961), pp. 69–85.

## What Is Contagion?

I shall now proceed to discuss Contagion, and shall begin with what seem to be its universal principles from which are derived its particular causes. . . . As its name indicates, contagion is an infection that passes from one thing to another. . . . The infection is precisely similar in both the carrier and the receiver of the contagion; we say that contagion has occurred when a certain similar taint has affected them both. So, when persons die of drinking poison, we say perhaps that they were infected, but not that they suffered contagion; and in the case of things that naturally go bad when exposed to the air, such as milk, meat, etc., we say that they have become corrupt, but not that they have suffered contagion.

Everything that happens, whether actively or passively, affects the essential substance of bodies or their non-essential parts. When someone has been heated or sullied by something, we do not say, except by metaphor, that he has suffered a contagion; because contagion is precisely a similar infection of the actual cause. Now when a house catches fire from the burning of a neighboring house, are we to call that contagion? No, certainly not, nor in general when a whole thing is destroyed primarily as a whole. The term is more correctly used when infection originates in very small imperceptible particles, and begins with them, as the word "infection" implies; for we use the term "infected," not of a something that is destroyed as a whole, but of a certain kind of destruction that affects its imperceptible particles. By the whole, I mean the actual composite, and by very small, imperceptible particles, I mean the particles of which the composite and mixture (combination) are composed. Now burning acts on the thing as a whole, whereas contagion acts on the component particles, though by them the whole thing itself may presently be corrupted and destroyed.

Contagion, then, seems to be a certain passive effect of elements in combination. But since such combinations can be corrupted and destroyed in two ways, either by the advent of a contrary element, owing to which the combination cannot retain its form or secondly by the dissolution of the combination, so happens when things have putrefied, we may perhaps hesitate to say whether contagion, when it is carried by the infection of the smallest particles, is produced in the former way or the latter. Moreover, what shall we say is the nature of this infection? Is it a corruption of those particles, or only an alteration? What, in short, happens to those particles? Hence it is hard to determine whether every contagion is a kind of putre-faction. All these problems will become clearer if we first investigate the fundamental differences of contagions and their causes. Meanwhile, if we allow ourselves to sketch a sort of tentative definition of contagion, we shall define it as: A certain precisely similar corruption which develops in the substance of a combination, passes from one thing to another, and is originally caused by infection of the imperceptible particles.

## The Fundamental Difference in Contagions

There are, it seems, three fundamentally different types of contagion: the first infects by direct contact only; the second does the same, but in addition, leaves fomes, and this contagion may spread by means of that fomes, for instance scabies, phthisis [tuberculosis], bald spots, elephantiasis, and the like. By fomes, I mean clothes, wooden objects, and things of that sort, which though not themselves corrupted can, nevertheless, preserve the original germs of the contagion and infect by means of those; thirdly, there is a kind of contagion which is transmitted not only by direct contact or by fomes as intermediary, but also infects at a distance; for example, pestilent fevers, phthisis, certain kinds of ophthalmia, exanthemata of the kind called variolae [small pox], and the like. These different contagions seem to obey a certain law; for those which carry contagion to a distant object infect both by direct contact and by fomes; those that are contagious by means of fomes are equally so by direct contact; not all of them are contagious at a distance, but all are contagious by direct contact. Hence the most simple kind of contagion is that of direct contact only, and it is naturally the first in order. . . .

## Contagion That Infects by Contact Only

An especially good instance of the contagion that infects by contact only is that which occurs in fruits, as when grape infects grape, or apple infects apple; so we must try to discover the principle of this infection. It is evident that they are infected because they touch; and that one fruit decays first, but what is the principle of the infection? Since the first fruit from which all infection passes to the rest has putrefied, we must suppose that the second has contracted a precisely similar putrefaction, seeing that we defined contagion as a precisely similar infection of one thing by another. Now putrefaction is a sort of dissolution of a combination due to evaporation of the innate warmth and moisture. The principle of that evaporation is always foreign heat, whether that heat be in the air or in the surrounding moisture; hence, in both fruits, the principle of contagion will be the same as the principle of putrefaction, namely extraneous heat; but this heat came to the first fruit either from the air or some other source; and we may not yet speak of contagion; but the heat has passed on to the second fruit by means of those imperceptible particles that evaporate from the first fruit, and now there is contagion, since there is a similar infection in both fruits; the heat that evaporates from the first fruit has power to produce in the second fruit what the air produced in the first, and to make it putrefy in a similar way, all the more because there is analogy. Now some of the particles that evaporate from the first fruit are hot and dry, either independently or when in combination. Those that are hot and dry are more apt to burn the fruit, whereas those that are hot and moist are more apt to produce putrefaction and less apt to burn. For the moisture softens and relaxes the parts of the fruit that it touches, and makes them easily separable, while the heat lifts them up and separates them. Hence when heat and moisture are produced within and evaporate, the result is dissolution of the combination, and this was our definition of putrefaction. We must therefore suppose that the hot moist particles—moist either independently or in combination—that evaporate from the first fruit, are the principle and germ of the putrefaction that occurs in the second fruit. I use the term "moist in combination," because, in evaporations that occur in putrescent bodies, it nearly always happens that very small particles are intermingled, and thus become principles of certain generations and of new corruptions; and this combination of hot and moist particles is most apt to convey putrefaction and contagions. We must therefore suppose that it is by means of these principles that contagion occurs in fruits. But in all other bodies also that are in contact and putrefy, if they are analogous to one another, it is reasonable to suppose that the same thing happens, and by means of the same principle. Now the principle is those imperceptible particles, which are hot and sharp when they evaporate, but are moist in combination. In what follows, they are called Germs of Contagion.

### Contagion That Infects by Fomes

Now it is at once obvious that the germs that transmit contagion by means of fomes are produced in the same manner and by the same principle as that above described, for the principle that exists in fomes seems to be of a different nature, inasmuch as, when it has retired into the fomes from the body originally infected, it may last there for a very long time without any alteration. Things that have been touched by persons suffering from phthisis or the plague are the most amazing examples of this. I have often observed that in them this virus has been preserved for two or three years; whereas particles that evaporate from putrefying bodies never seem to have the power to last as long as that. Nevertheless no one ought to think that the principle of contagion that is in the fomes is not the same as the principles that infect by contact only, because the very same particles that evaporate from the body originally infected, after being thus preserved, can produce the same effect as they would have done when they evaporated from the original body. . . . Now a combination is strong and lasting in virtue of two qualities; first, it must have the kind of hardness possessed by iron, stones, and the like, whose very small, imperceptible particles last for many years; secondly there must be present a certain viscosity, and the mixing process must be thoroughly elaborated. So that even when the germs of contagion are not hard, they be viscous and elaborated. By an elaborated combination I mean one composed of very small particles well shaken together. . . . Combinations of this sort are produced by evaporations that are closely confined, where what evaporates is not dispersed, but is violently shaken, and hence is very finely and minutely mixed. Now if viscosity be added, the resulting combination is strong and suitable for its preservation in fomes. A proof of this is that all germs that infect by means of fomes are without exception viscous and sticky, and only when they have this quality can they occupy fomes. . . .

## Contagion at a Distance

Even more surprising and hard to explain are those diseases that cause contagion, not by direct contact only or by fomes only, but also at a distance. There is a kind of ophthalmia with which the sufferer infects everyone who looks at him. It is well-known that pestiferous fevers, phthisis, and many other diseases infect those who live with the sufferer, even though there is no actual contact. It is far from certain what is the nature of these diseases, and how the taint is propagated. We must therefore study these problems with the greatest care, since of this sort are the majority of the diseases we are investigating.

## How the Germs of Contagion Are Carried to a Distant Object and in a Circle

Let us first enquire by what sort of movement these germs of contagion are impelled, since it is clear that they are carried far and to persons far distant, a fact that many people find astonishing. . . . Now the principle of movement in these small bodies, in all directions, is in part independent, in part given by something else. All the evaporation independently rises upwards, as may be seen in smoke and many other things, for everyone knows that all evaporation is warm; but the movement may be derived from something else, and then the thrust is sideways and finally downwards. This is due to two main causes, one being the resistance of the air or of the floors, the things of that sort on which the particles that are first exhaled; when these particles cannot be carried further, they are thrust sideways by the particles that follow them, and these by others, till the whole surrounding space is filled. The second cause is the air itself which divides into its small and indivisible parts all evaporation that is tenuous and easily soluble. For it is the nature of elements and of all liquids that they seek, so far as it is possible, a suitable position; and a position is most suitable when the parts are continuous, or if not continuous, are the least possible distance apart from one another; for thus they are less exposed to violence. . . . Hence the air keeps on dividing the evaporation more and more, until it arrives at those parts which cannot be further divided and separated. Then when this countless division has been made, much of the air is filled and mingled with the evaporation all round and about, as is most evident in the case of smoke. These then are the reasons why the evaporations that occur in contagions are also carried around and about, and occupy a great volume in the air. . . .

Thus it is that these germs may infect also those who live with persons infected, and the germs can be preserved for a certain time, not only in fomes but also in the air, though no longer in fomes.

But how does it happen that germs whose bulk is so small do not suffer alteration, when thus exposed to the air? That is the first question. What must be the strength of the combination in so small a particle, especially when those particles have not the quality of hardness that they can last so long in the air? It is those that are viscous and sticky. However small that they may be, they can live, if not quite so long as the hard ones, still nearly as long.

The hard particles offer most resistance to alterations, because of three properties. First, in a small bulk they have more substance; secondly, they are colder, on account of their earthy elements; thirdly, by reason of their density, their parts cannot be easily volatilised and rarefied as ought to happen when heat is introduced. . . . [H]ard but also viscous bodies, defend themselves from many alterations, if only these are moderate; but they cannot endure violent alterations. Hence the germs of all contagions are consumed by fire, and are broken up by very cold water also. . . .

Of such sort are the germs of contagions also, for they are all per se acute, although constituted in viscosity, and they become active when the animal heat vaporises that combination and brings together the similar parts. Now germs of this kind have great power over the humors and spirits, so that they can even cause death in a few hours if they are analogous with the spirits, but about this I will say more later. These same germs can be shot forth from sore eyes into the eyes of another and carry in a precisely similar infection, and this is not a visual image but a taint in the eye. It is not surprising, if one considers the method by which they attack, that they penetrate into the animal, and some of them very quickly, for they attack and enter from the small pores, veins, and arteries into the larger, and from these to others and often reach to the heart.

One method of penetration is by propagation, and so to speak, progeny. For the original germs, which have adhered to the neighboring humors with which they are analogous, generate and propagate other germs precisely like themselves, and these in turn propagate others, until the whole mass and bulk of humors is infected by them. A second method of penetration is by attraction, which works inwardly, partly through the breath of inspiration, partly by the dilation of blood vessels. For along with the air that is drawn in, there enter, mixed with it, germs of contagions, and when once these have been introduced, they do not retire as easily by expiration as they entered by inspiration; for they adhere closely to the humors and organs, and some of them even to the spirits, which retreat from the image of their contrary, and carry their enemy with them even to the heart. . . .

## The Analogy of Contagions

Contagions have manifold and very surprising analogies [selective properties]. For instance, there is a certain pest which attacks trees and crops, but harms no sort of animal; again there is a pest which attacks certain animals but spares trees and crops. In the animal world, one pest will attack man, another cattle, another horses and so on. . . . Some pests work promiscuously, so that some persons are infected, others not; some persons can associate with the plague-stricken and take no hurt, others can not. The organs of the body also have their own analogy (affinity), for ophthalmia harms no organ save the eyes, while phthisis does not affect the eyes, though they are so delicate, but does affect the lungs. . . .

## Is Every Contagion a Kind of Putrefaction?

Now let us enquire whether every kind of contagion is a kind of putrefaction, and whether every putrefaction is contagious. It seems that every putrefaction is contagious, either absolutely, or at least contagious to a contiguous part, but not every putrefaction is contagious for another body, since, in order that it may act, many factors are required, as I have said. Perhaps one may doubt whether every contagion consists in putrefaction of some sort; since rabies seems to be contagion of a sort, but not putrefaction. Likewise, when wine turns to vinegar, it seems to suffer a sort of contagion by something else, but not to suffer putrefaction. For when it putrefies it has a bad smell and is unfit to drink, whereas vinegar is pleasant to taste and even opposes putrefaction. Yet these cases must also be regarded as putrefactions of a sort. . . .

Now all putrefactions have the power to convey precisely similar putrefaction, at least to a continuous part; hence, if every contagion is putrefaction, it seems that contagion, simply and generally speaking, might be defined as: A certain precisely similar putrefaction which passes from one thing to another, whether that other be continuous with the original thing or separated from it. Yet this is not contagion strictly so called, for true contagion occurs between two different bodies. But if we wish to consider, above all and by itself, that contagion which is observed in diseases and does not affect by direct contact only, then we shall define contagion as: A precisely similar putrefaction which passes from one thing to another; its germs have great activity; they are made up of a strong and viscous combination; and they have not only material but also a spiritual antipathy to the animal organism. This definition will give us the key to all the phenomena that are observed in contagion.

### Exercises

**1. Definition**
   Look up the word *germ*. Where does Fracastoro get the term? Write an extended definition of this term. Go beyond this essay in your definition.

## 2. Stipulative Definition

Write a sentence definition for *contagion*, and use that as the basis to write a stipulative definition of the term. For this assignment, you should stay entirely within the text of this essay.

## 3. Comparison/Contrast

Write an essay that compares/contrasts infection and putrefaction. You may do further research into the historical background of these terms.

## 4. Analysis

Using both Hippocrates' and Thucydides' writing as a point of contrast, discuss the significance of a germ theory of disease.

# A *Contagium vivum fluidum* as the Cause of the Mosaic Diseases of Tobacco Leaves[1]

## Martinus W. Beijerinck

### (1899)

The general term for a living infectious agent which had been used for years was "virus." Any infectious agent was described with this word. Although many infectious diseases were easily shown to be caused by bacteria, it was not always possible to do so. Sometimes it was possible to transmit a disease from host to host even after the infectious fluid was filtered through pores so small that all bacteria were removed. When it could be shown that this agent could pass through a filter, then reproduce in a new host, and, after being filtered again, infect another host, it was obvious that here was a living agent. Beijerinck called this a "liquid" (or fluid, that is non-particulate) living agent of disease. Others called these "Filterable viruses." Gradually the word "virus" became restricted to filterable viruses, so that now that adjective filterable has been dropped. We know now that these viruses are particulate, and since [Wendell] Stanley succeeded in crystallizing one, we know that they are quite unlike other living organisms, so that many refused to consider them living. —Thomas Brock, editor and translator, *Milestones in Microbiology* (Washington, D.C.: American Society for Microbiology, 1961), p. 6.

Although Beijerinck was wrong about the diffusibility of the tobacco mosaic virus, the rest of his observations are quite valid. He describes an agent which can pass through the smallest filters, can apparently reproduce only in the living plant and seems quite stable. In attempting to explain these observations, he finds himself in a dilemma since the physiological and biochemical facts of cell function were not yet available for him to use in explaining his observations. It is interesting that he comes as close to hitting the nail on the head as he does. His postulate that the virus becomes incorporated through living protoplasm of the host plant is one which is about as close to current thinking on virus multiplication as it would be possible to get in 1899. . . . He shows that the amount of virus

227

in a filtrate could be crudely quantitated. This is an important area in virus research, since it is necessary to have some idea of how much infectious material is present in a sample. (Comment on Beijerinck, *Milestones in Microbiology*, p. 157.)

---

The leaf spot disease of tobacco, also called the mosaic disease, is manifested first as a bleaching of the chlorophyll, occurring in spots over the leaf blade. This is followed later by the death of a part or all of the tissue of the spots. The discoloration appears first right next to the leaf veins and it is manifested then by a strong increase in the amount of chlorophyll. Later the spaces between the spots become bleached usually to a yellow color, but in isolated cases, they become completely albino. The dark green patches grow at the beginning more rapidly than the other parts of the leaf, leading to wart-like growths which arise from the upper surface of the leaf. However, this phenomenon is observed more often in artificial infections than in tobacco fields, where the diseased leaves usually remain completely flat.

The third phase of the disease consists of localized death of the hundred or thousand small spores which are distributed randomly over the leaf. These then assume a brown color and become very fragile, so that the holes are formed easily during the harvest of the leaves. These spots are the fear of the Dutch tobacco farmer, because they make the leaf worthless for cigar wrappers. . . .

Herr Adolf Mayer showed in 1887 that this disease was contagious. He expressed the sap from sick plants, placed it in capillary tubes and stuck these in healthy plants. He found that after 2–3 weeks, the latter plants became diseased.

In 1887 I attempted to discover if there was not a parasite which could be demonstrated to be the cause of the disease. Since microscopic studies were completely negative, the only type of bacteria that could be considered were those which could not be observed directly. But culture procedures showed that aerobic bacteria were completely absent, either from the healthy or the diseased plants. I later showed that anaerobic bacteria were also absent.

It seemed certain, therefore, that we were dealing here with a disease which was caused by a *contagium* which was not a *contagium fixum* in the usual sense of the words. This encouraged me to carry out new experimental infections in 1897 and 1898, in order to understand the properties of the *contagium* better. I would like to present here briefly the main results which were obtained from these studies.

It was first shown that the juice expressed from sick plants did not lose its virulence even after being filtered through a porcelain filter so fine that it rendered the juice completely sterile. This filtrate was tested for the presence of both aerobes and anaerobes, so that the experiment was completely unobjectionable. This filtrate was kept three months and remained completely bacterial free during this time but was repeatedly shown to induce the identical mosaic when inoculated into plants. I do not know how long the virulence of this filtrate can be maintained.

The following experiments were designed to answer the question as to whether the virus should be considered particulate or soluble.

Pulverized tissue of diseased leaves was spread on thick agar plates and diffusion allowed to occur. A virus which was particulate would remain on the surface of the agar since it could not diffuse into the molecule-sized pores of the agar plate. The deep layers of the agar would thereupon become virulent. But a water soluble virus ought to be able to penetrate to a certain depth in the agar plate. The experiment was discontinued after a diffusion time of about ten days, what would be considered to be long enough, since I knew that diptase trypsin would diffuse in a considerable extent in this time. The upper surface of the plate was first washed with water and then with a strong solution of mercuric bichloride. After this, a sharp platinum needle was useful to remove part of the agar, so that the inner layers could be reached, care being taken not to disturb the upper surface. Healthy plants were then infected with agar from these deep layers. The infection was just as extensive with this material as when the sterile filtrate was used. It can hardly be doubted, therefore, that the contagium must be considered to be fluid, or more accurately, water soluble.

The experimental infections using plant juices were performed using the hypodermic needle of Pravaz. The most suitable place to infect is the youngest part of the stem which can be manipulated easily without causing extensive damage, since the closer the infection is to the meristem of the terminal bud, the earlier the results are seen. It has been shown that the virus moves slowly through the plant, and further, that only the portions of the young leaves that are undergoing cell division are sensitive to the infection. Both the mature leaves as well as the young leaves in which the cells have already stopped dividing are completely insensitive to the virus, even though they are able to transport it towards the meristematic regions. If stem internodes that are enlarging are infected, after 10–12 days the first symptoms of the disease can be observed in the young leaves which are coming out of the apical meristem. However, if an infection is carefully made as close as possible to the apical meristem, even after 3–4 days, yellow spots and crisp distorted areas can be observed in the youngest little leaves that are still within the bud.

The amount of virus which is sufficient to infect a large number of leaves is quite small. It is then possible to obtain material from these diseased leaves which can be used to infect unlimited numbers of new plants. It is therefore quite clear that the virus is reproducing within the plant. From the above, it is clear that this reproduction is not in the mature plant cells but in those tissues where cell division is occurring.

Although the virus can exist outside the tobacco plant, it cannot reproduce under these conditions. I conclude this from the following fact: If a sterile filtrate of virus is mixed with a healthy tobacco plants, it can be determined by experimental infections that no reproduction of the virus is obtained. Instead the virus is diluted in the same way as if pure water had been used instead of plant sap.

It is not difficult to determine the accuracy of this statement, since the amount of virus used to infect plants has been a great influence on the development of the symptoms of the mosaic disease. With a small amount of virus, the usual results are obtained as described above. With large amounts of virus, highly-deformed leaves of characteristic shapes are obtained. In order to obtain these deformed leaves, it is necessary to inject much more of a diluted virus than of one not diluted. In this way, it is easy to tell whether the virus has reproduced or stayed the same in any type of fluid. As mentioned above, I have not observed reproduction under artificial conditions, so that I believe the only model of reproduction of the virus is in the cells of the plants that are dividing.

The ability of the virus to reproduce only when combined with living protoplasm of the host plant maybe related to its soluble or liquid nature. It is not easy to understand why a *contagium fixum*, even if so small that it could not be seen by direct microscopic examination, could not still reproduce away from the host, like ordinary parasitic bacteria. In addition, it would seem probable that a microscopically invisible *contagium*, if particulate, could develop into macroscopically visible colonies on gelatin plates.

A soluble and diffusible virus, such as the mosaic virus, should bring about some coloration or change in refractive index of a gelatin or sugar medium, if the chemical nature of the medium were altered when used as a nutrient by a reproducing virus. Such changes could not be seen when the virus was seeded onto malt extract gelatin or onto plates containing 10 per cent gelatin dissolved in a plant decoction containing 2 per cent cane-sugar, both excellent media in my hands for the growth of parasitic and saprophytic plant bacteria. It also seems to me that reproduction or growth of a soluble body is not inconceivable, although difficult to imagine. It would not seem wise to assume a division process of molecules which would lead to their reproduction, and the idea of molecules which feed themselves, which much be assumed to explain this, seems to me an unclear concept, if not actually contrary to nature.

A partial explanation would be the view that the *contagium* must be incorporated into the living protoplasm of the cell in order to reproduce, and its reproduction is so to speak passively brought about with the reproduction of the cell. But this would then leave us with one mystery instead of two, since the incorporation of a virus into living protoplasm, even if it is shown to be a fact, can in no way be viewed as an understandable process.

If the soil in which a tobacco plant is growing is infected with the virus, after a time the disease is seen to appear in the apical bud. The length of time for its appearance is primarily dependent on the size of the plant. In young plants, I saw the first symptoms in two weeks, while in larger and older individuals, 4–6 weeks occurred before the symptoms appeared in the newly formed leaves of the terminal meristem. Therefore the roots and stem must be able to transmit the virus considerable distances. . . .

It is possible to infect the plants through the roots only when they are two or more decimeters high. It is uncertain whether wounds in the roots are necessary, or whether the uptake of the virus can occur through surface contact at the root. Since the *contagium* can only attack the leaves that form after the infection begins, the number of healthy leaves below the infected ones can be used to approximate the time of infection in plants growing naturally which have taken up the virus through their roots.

The virus can be dried without a change in its virulence. It could therefore overwinter in soil, where it perhaps would be partially destroyed like so many bacteria and yeast.

An alcohol precipitate of virulent plant juice, dried at 40 degrees C., retained its virulence.

The virulence is also maintained in dried leaves, so that two-year old herbarium leaves are still suitable for experimental infections. Therefore the dried dust which forms easily during the harvest from the broken dead tissue of the leaf spots must undoubtedly be able to spread the disease.

As expected in a moist environment, the virus was inactivated by boiling water, as well as at 90 degrees C. I have not determined the lowest temperature at which activation would occur but would expect it would between 70 and 80 degrees C. . . .

It is possible that there are a whole series of plant diseases which are caused by *contagium fluidum*, in a similar manner to the mosaic disease of tobacco plants. The diseases of peach trees described in America by Erwin Smith sometime in 1894 under the names peach yellows and peach rosette seem, from his description, undoubtedly to belong here, although it is not yet certain if these diseases can be transmitted only through budding and grafting, as he describes, or, what is more likely, they can also be transmitted through the juice of the dead tissues.

## Exercises

1. Describe the phases of the tobacco mosaic disease.
2. Though not phrased as a question, define the research question which directs the formulation of Beijerinck's hypotheses.
3. Identify the hypotheses.
4. Describe the series of experiments. Distinguish between Beijerinck's descriptions of past research and present research. Make a clear statement about how the experiments are connected as whole.

# Molecular Structure of Nucleic Acids
## A Structure for Deoxyribose Nucleic Acid[*]

J. D. Watson and F. H. Crick

We wish to suggest a structure for the salt of deoxyribose nucleic acid (D.N.A). This structure has novel features which are of considerable biological interest.

A structure for nucleic acid has already been proposed by Pauling and Corey.[1] They kindly made their manuscript available to us in advance of publication. Their model consists of three intertwined chains, with the phosphates near the fibre axis, and the bases on the outside. In our opinion, this structure is unsatisfactory for two reasons: (1) We believe the material which gives the X-ray diagrams is the salt, not the free acid. Without the acidic hydrogen atoms it is not clear what forces hold the structure together, especially as the negatively charged phosphates near the axis will repel each other. (2) Some of the van der Waals distances appear to be too small.

Another three-chain structure has been suggested by Fraser (in the press). In his model, the phosphates are on the outside and the bases on the inside, linked together by hydrogen bonds. This structure as described is rather ill-defined, and for this reason, we shall not comment on it.

We wish to put forward a radically different structure of the salt of deoxyribose nucleic acid. This structure has two round helical chains each coiled round the same axis (see diagram).

We have made the usual chemical assumptions, namely, that each chain consists of phosphate diester groups joining β-D-deoxyribofuranose residues

---

[*] *Nature* 171, No. 4356 (25 April 1953): 737–38. *Reprinted by permission from Nature. Copyright © 1953 Macmillan Magazines Ltd.*

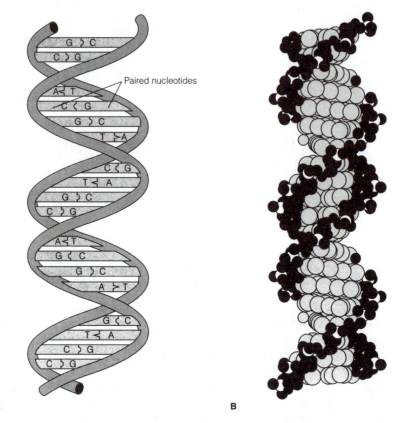

Paired nucleotides

A                                                    B

The DNA molecule: (A) diagrammatic model of the double helix showing the pairing of A in one strand with T in the other and of G in one strand with C in the other; (b) space-filling model of the DNA double helix. Each DNA strand winds around the other with the base pairs bridging between the strands.

with 3′, 5′ linkages. These two chains (but not their bases) are related by a dyad perpendicular to the fibre axis. Both chains follow right-handed helices, but owing to the dyad, the sequences of the atoms in the two chains run in opposite directions. Each chain loosely resembles Furnberg's[2] model No. 1; that is, the bases are on the inside of the helix and the phosphates on the outside. The configuration of the sugar and the atoms near it close to Furberg's 'standard configuration,' the sugar being roughly perpendicular to the attached base. There is a residue on every chain every 3-4 Å in the z-direction. We have assumed an angle of 36° between adjacent residues in the same chain, so that the structure repeats itself after 10 residues on each chain, that is, after 34 Å. The distance of a phosphorus atom from the fibre axis is 10 Å. As the phosphates are on the outside, cations have easy access to them.

The structure is an open one, and its water content is rather high. At lower water contents, we would expect bases to tilt so that the structure could become more compact.

The novel feature of the structure is the manner in which the two chains are held together by purine and pyrimidine bases. The places of the bases are perpendicular to the fibre axis. They are joined together in pairs, a single base from one chain being hydrogen-bonded to a single base from the other chain, so that the two lie side by side with identical z-co-ordinates. One of the pair must be a purine and the other a pyrimidine for bonding to occur. The hydrogen bonds are made as follows: purine position 1 to pyrimidine position 1; purine position 6 to pyrimidine position 6.

If it assumed that the bases only occur in the structure in the most plausible tautomeric forms (that is, with the keto rather than the enol configurations) it is found that only specific pairs of bases can be bound together. These pairs are : adenine (purine) with thymine (pyrimidine), and guanine (purine) with cytosine (pyrimidine).

In other words, if an adenine forms one member of a pair, on either chain, then on these assumptions the other member must be thymine; similarly for a guanine and cytosine. The sequence of bases on a single chain does not appear to be restricted in any way. However, if only specific pairs of bases can be formed, it follows that if the sequence of bases on one chain is given, then the sequence on the other chain is automatically determined.

It has been found experimentally[3,4] that the ratio of the amounts of adenine to thymine, and the ration of guanine to cytosine, are always very close to unity for deoxyribose nucleic acid.

It is probably impossible to build this structure with a ribose sugar in place of the deoxyribose, as the extra oxygen atom would make too close a van der Waals contact.

The previously published X-ray data[5,6] on deoxyribose nucleic acid are insufficient for a rigorous test of our structure. So far as we can tell, it is roughly compatible with the experimental data, but it must be regarded as unproved until it has been checked against more exact results. Some of these are given in the following communications. We were not aware of the details of the results presented there when we devised our structure, which rests mainly though not entirely on published experimental data and stereochemical arguments.

It has not escaped our notice that the specific pairing we have postulated immediately suggests a possible copying mechanism for the genetic material.

Full details of the structure, including the conditions assumed in building it, together with a set of co-ordinates for the atoms, will be published elsewhere.

We are much indebted to Dr. Jerry Donohue for constant advice and criticism, especially on interatomic distances. We have also been stimulated by a knowledge of the general nature of the unpublished experimental results and ideas of Dr. M. H. F. Wilkins, Dr. R. E. Franklin, and their co-workers at King's College, London. One of us (J.D.W.) has been aided by a fellowship from the National Foundation for Infantile Paralysis.

*J. D. Watson and F. H. Crick*
*Medical Research Council Unit for the Study of*
*the Molecular Structure of Biological Systems*
*Cavendish Laboratory*
*Cambridge*
*April 2*

## Notes

1. Pauling, L., and Corey, R. B., *Nature* 171 (1953): 346; *Proc. U.S. Nat. Acad. Sci.* 39, 84 (1953).
2. Furberg, S., *Acta Chem. Scand.* 6, 634 (1952).
3. Chargaff, E., for reference see Zamenhof, S., Brawerman, G., and Chargaff, E., *Biochem. et Biophys. Acta* 9, 402 (1952).
4. Wyatt, G. R., *J. Gen. Physiol.* 36, 201 (1952).
5. Astbury, W. T., *Symp. Soc. Exp. Biol.* vol. 1: *Nucleic Acid* 66 (Cambridge University Press, 1947).
6. Wilkins, M. H. F., and Randall, J. T., *Biochem et Biophys. Acta* 10, 192 (1953).

## Exercises

### 1. Order of Description
Analyze the ordering of the description of the DNA molecule.

### 2. Significance
This essay was one of the most important of our time, yet it does not announce its own importance with a great deal of fanfare. How could you know the significance of this information? Hint: analyze this paper for its targeted audience.

### 3. Scientific Method
Analyze Crick and Watson's model for its explanatory value. Is this model presented hypothetically? Must further testing be conducted to validate its efficacy? Locate this paper in the process of scientific reasoning.

# Appendices

# Appendix I

Writing Grant Proposals

A grant proposal is a formal, written request for money, equipment, library, or computer access. Whether proposals run to a few pages or to a hundred pages, they require a statement of the request, an explanation/justification for the request, a budget, and a specific plan of implementation. As a written form, the grant is explicit and straightforward, but grant-writing is time consuming and labor intensive because in addition to describing your project in a written text, you must fill out forms, devise a budget, and in many cases collect the curricula vitae of the members of the research team. If your budget lists travel costs, you will have to call airlines for ticket estimates or consult sources for mileage. If the budget lists equipment costs, you will have to solicit estimates. These logistical efforts require time and organization.

Budget your time, not only to accommodate the logistics, but also to allow for several revisions of the proposal. Soliciting readings by your colleagues will provide you valuable feedback on all aspects. Soliciting a reading from a person outside the field can provide you with feedback on the general intelligibility of your grant. Published deadlines—*postmark* deadlines or *receipt-of-materials*—are taken seriously. *Check your deadline.*

In the sciences, a proposal will be composed of the following parts: an introduction, abstract, statement of the research problem, methodology used to implement the research program, and anticipated results. In structure, a proposal is analogous to a science paper with anticipated results taking the place of actual results. Some granting institutions will ask you to submit a schedule for carrying out the research program.

If you are asked to submit a schedule of research, and contingencies make you tentative about timing, estimate. Put some thought into planning this. When it comes time to actually implement your research plan, you may find yourself turning to those timing guidelines in order to fulfill them.

## Evaluation Criteria

Granting agencies make explicit the criteria by which a grant is evaluated.

No privileged information is withheld from you. As you formulate the argument for your proposal, write with the evaluation criteria in mind. Because proposal format and evaluation criteria differ, make sure that you respond to the specific requirements of the granting institution.

A science proposal is evaluated roughly on two sets of criteria: (1) scientific merit and (2) implementation. By scientific merit is meant that the solution to your research problem should produce a substantive contribution to your area of study or to the public. Some granting agencies look more favorably upon projects which engage in applied science research rather than theoretical research. Some agencies may value applied and theoretical work equally. Make sure that your project is directed to the institution which funds the kind of research you propose.

The grant is assessed by a number of factors. Your ability to carry out the goals of the proposal is determined by your experience and that of your staff, evaluated in part through the curricula vitae you are asked to supply. Your institution's ability to support the research is also a factor weighed by the review board; if you need a supercollider to conduct research at a facility which has no supercollider, your proposal is obviously at a disadvantage. In addition, the project must be planned and implemented to conform with your institution's policy and procedures regulations. Those regulations may limit the use or nature of research subjects or materials; they may restrict the way you spend grant funds. The research must conform to national policies or protocols regulating the use of subjects, materials and so on.

The following extract from the National Science Foundation, Grant Proposal Guide [(NSF 94-2), January 1994] defines the selection criteria for grants.

> Four criteria for the selection of research projects by the National Science Foundation are listed below, together with the elements that constitute each criterion.
>
> **(1) Research performance competence**—This criterion relates to the capability of the investigator(s), the soundness of the proposed approach, and the adequacy of the institutional resources available.
>
> **(2) Intrinsic merit of the research**—This criterion is used to assess the likelihood that the research will lead to new discoveries or significant advances within the field of science or engineering, or have a substantial impact on progress in that field or in other scientific and engineering fields.
>
> **(3) Utility or relevance of the research**—This criterion is used to assess the likelihood that the research can contribute to achievement of a goal that is extrinsic or in addition to that of the research field itself, and thereby serve as the basis for new or improved technology or assist in the solution of society problems.
>
> **(4) Effect of the research on the infrastructure of science and engineering**—This criterion relates to the potential of the proposed research to contribute to the better understanding or improvement of the quality,

distribution, or effectiveness of the Nation's scientific and engineering research, education, and manpower base.

## The Review Process

Your grant may pass through several levels of review. When your grant is received by the granting institution, it may be sent to readers knowledgeable in the field. In some cases, the reviewer may offer recommendations without convening as a group. The NIH (National Institutes of Health) for example has a two-tiered review process, the members of which serve only in an advisory capacity; they make recommendations but do not ultimately make decisions. In the first tier, the proposal is sent to experts in the field who assign a priority to your work based on scientific merit; they do not, however, set policy.

The second tier of review goes over the first-tier recommendations and assesses the applications against the program's priorities; this level of review is composed of members who are inside and outside the science field. Ultimately decisions rest with a third group, the Division of Research Grants (DRG), which considers the recommendations made by the tiers.

The tiered review process puts your proposal among diverse readers. To which audience do you write? Your writing must be intelligible to all your readers and, as such, must be concise, clear, and to the point. Avoid overly technical language. Bear in mind that reviewers have other proposals besides yours to read in addition to the daily concerns of working, raising a family, or maintaining their own professional profiles. Do not make your reader work too hard. Your proposal should stand out by its merit, by the soundness of its procedural design, and by the clarity of its writing.

The review board at any level makes one of three decisions: approved, declined, deferred. A deferral is granted to those proposers who need to provide further information before action can be taken. Approval means that the board has found your proposal satisfactory in scientific merit and implementation strategies. Even though your proposal may secure approval at one or two levels in the tier-process, it still remains in competition with other grants which have likewise secured approval and will hence be ranked against them. If the proposal is declined, it is knocked out outright. Deferral means that a decision has been put off pending the submission of further materials by the applicant or pending a site visit. The purpose of a site visit is to assess the capabilities of the institution to implement the research project.

## Budgets

Budgets are usually accompanied with a narrative which explains the costs.

Use headings if your budget narrative has many subsections. In planning travel, solicit estimates for business class. If your grant originates under the auspices of your university, your university will probably prohibit such items as first-class travel, monies for entertainment and liquor. Direct costs refer to the actual cost of the project; indirect costs, usually expressed as a percentage, refer to the grant funds which go to the institution to cover overhead costs like electricity, laboratory or office space, and equipment. Indirect costs may run as high as 50 percent of the grant award.

# Headings

Use headings to divide the parts of your proposal. The grant specifications spell out particular divisions. Derive headings from these categories; use sub-headings as necessary.

### Proposal Specification: NIH

The following specifications are taken from the application package for a Public Health Service Grant issued by the U. S. Department of Health and Human Services, Grant Application form PHS 398 and explanations (6/30/94). They are intended to illustrate the prescriptive nature of grant proposals.

**Research Plan.** Organize Items 1-4, to answer these questions: (1) What do you intend to do? (2) Why is the work important? (3) What has already been done? (4) How are you going to do the work? **Do not exceed 25 pages for Items 1-4.** All tables and graphs must be included within the 25 page limit of Items 1-4. Full-sized glossy photographs of material such as electron micrographs or gels may be included in the Appendix; however, a copy of any photograph must also be included within the Research Plan and within the page limitations (See Appendix for further instructions). **Twenty five pages is the absolute maximum and will be strictly enforced. Applications that exceed this limit, or that exceed the type size limitations (see page 10), will be returned without review.** You may use any page distribution within this overall limitation; however, the PHS recommends the following format and distribution:

1. **Specific Aims.** State the broad, long-term objectives and describe concisely and realistically what the specific research described in this application is intended to accomplish and any hypotheses to be tested. **One page is recommended.**

2. **Background and Significance.** Briefly sketch the background to the present proposal, critically evaluate existing knowledge, and specifically identify the gaps which the project is intended to fill. State concisely the importance of the research described in this application by relating the specific aims to the broad, long-term objective. **Two to three pages are recommended.**

3. **Progress Report/Preliminary Studies.** A progress report is required for **Competing Continuation** and **Supplemental** applications; for

**New** applications, a report of the principle investigator/program director's preliminary studies is recommended.

For COMPETING CONTINUATION and SUPPLEMENTAL applications, give the beginning and ending dates for the period covered since the project was last reviewed competitively. List **all** personnel who have worked on the project during this period, their titles, birth dates, Social Security Numbers, dates of service and percentages of their appointments devoted to this project. Summarize the previous application's specific aims and provide a succinct account of published and unpublished results indicating the progress toward their achievement. Summarize the importance of the findings. Discuss any changes in the specific aims since the project was reviewed competitively. List the titles and complete references to all publications, manuscripts **submitted or accepted** for publication, patents, invention reports, and other printed materials that have resulted from the project since it was last reviewed competitively. **Note that this list is excluded from the 25 page limit.** Submit five collated sets of **no more than ten** such items as an Appendix.

NEW applications may use this section to provide an account of the principle investigator/program director's preliminary studies pertinent to the application and/or any other information that will help to establish the experience and competence of the investigator to pursue the proposed project. The titles and complete references to appropriate publications and manuscripts **submitted or accepted** for publication may be listed, and five collated sets of **no more than ten** such items of background material may be submitted as an Appendix. **Six to eight pages are recommended for the narrative portion of the Progress Report/Preliminary Studies, excluding the list of materials resulting from the project since it was last reviewed competitively.**

4. **Research Design and Methods.** Outline the research design and the procedures to be used to accomplish the specific aims of the project. Include the means by which the data will be collected, analyzed, and interpreted. Describe any new methodology and its advantages over existing methodologies. Discuss the potential difficulties and limitations of the proposed procedures and alternative approaches to achieve the aims. Provide a tentative sequence or timetable for the investigation. Point out any procedures, situations, or materials that may be hazardous to personnel and the precautions to be exercised. **Although no specific number of pages is recommended for this section of the application, the total for Sections 1–4 may not exceed 25 pages, including all tables and figures.**

# Discussion

1. *Specific aims* asks for the definition of long-term objectives. (See Chapter 8, "Definition"), and a description of the goals of your research project. The discussion of *causality* in Chapter 12 may be useful to this section. The proposed research is posited as a cause for which you must anticipate effects. In what position of causality do you see your work? Do your long-term objectives represent your project as a remote cause able to influence or contribute to a general knowledge base? Do your long-term objectives represent your project as a proximate cause with direct and clear

cut ends? Will it *influence, contribute, modify* other aspects of science? Is your project the first step in a series of projects, perhaps the initial research in what may be your life's work?

2. The background survey parallels the background information which serves as an introduction to a science paper. Notice that you have two pages to represent both the major contributions to the field and whatever commentary you have on them. While your research on the background must be comprehensive, you are not in a position to mention every title. This means that you must *classify* your research into patterns. Assess what the general consensus is, who the most significant proponents of that view are, who the exceptions are. Mention the notable researchers.

This section sets up a comparison/contrast mode as a means to defining the significance of your project. By explaining the status of the field (definition), and explaining how your work fills in the gap (comparison/contrast), you arrive at an explanation as to why your work is important. This argument from definition was introduced in Chapter 10 under the heading "Essay Exam." A further demarcation of significance can be structured through a comparison/contrast between your immediate and long-term research objectives.

Make sure that your mention of long-term objectives as stated in section 1 corresponds to the long-term objectives you discuss in section 2.

Section 3 asks for the researcher's credentials. What information about the person will contribute to the credibility of the researcher's plan? Credentials are established by past publications or past experience which demonstrate a record of achievement and which display expertise in the area.

Section 4 calls for an outline of methodology. An outline refers to a more schematic discussion that you would normally supply in a science paper. Since experts will be among the review board, gear this segment more towards that audience.

## What to Do If Your Project Fails to Get Funded

Do not lose heart if you do not acquire all the support you requested or if your proposal is declined. If you receive a summary statement which represents the reviewer's remarks, read it carefully to assess the limitations of your proposal. Determine if the limitation to your grant came from unclear writing: if you didn't get your point across, while the original project had merit, your manner of presentation hindered its success. Rewrite to clarify and resubmit.

If your grant was declined because you did not provide a realistic plan of implementation, spend more time on fleshing out the plan and resubmit. If your science project is construed to be without scientific merit, reconceive your project.

## References

Reif-Lehrer, Liane. *Grant Application Writer's Handbook.* Jones and Bartlett, 1995.
_____. *Writing a Successful Grant Application.* Jones and Bartlett, 1989.

# Appendix II

## Mechanics and Style

### The Deep Six

The "Deep Six" are six grammar/punctuation errors. They are called the deep six because they are the most frequently made obfuscating errors; they *sink* your prose into meaningless oblivion, interfering with the meaning of the writing by making a point difficult or impossible to follow. While such obstacles to communication, if infrequent or few, may be as innocent as "white noise," more often they will confuse the reader about your point.

The "Deep Six" are not the only grammar/punctuation errors but they are the most common and therefore the most necessary to correct. Read the following descriptions. If you do not understand what the errors are or how you can fix them, turn to the appropriate section in the grammar unit which follows.

**1. Sentence Fragment:** A sentence fragment is a group of words written as a sentence that fail to function as a grammatically complete sentence. A sentence conveys a complete idea. It should have a subject and a finite verb in a grammatically independent predication.

> Fault: Me Tarzan–you Jane.
> Correction: I am Tarzan. You are Jane
>
> Fault: Being that he was late.
> Correction: He was late.
>
> See *verbals* and *clauses*.

**2. Comma Splice:** A comma splice results when two independent clauses are run together with a comma. Comma splices, like repeated sentence fragments, suggest that the writer is unable to recognize complete ideas. Two independent clauses can be joined in two ways: (1) with a comma *and* a coordinating conjunction (and, or, but, nor, for); (2) with a semicolon.

Alternatively, (3) two independent clauses could be written as two separate sentences, or (4) one independent clause could be subordinated to the other.

Fault: We canceled the picnic, it rained.

Corrections:
| | |
|---|---|
| *(1) comma + conjunction* | We canceled the picnic, and it rained. |
| *(2) semicolon* | We canceled the picnic; it rained. |
| *(3) two independent clauses* | We canceled the picnic. It rained. |
| *(4) subordination* | We canceled the picnic because it rained |

See *clauses.*

**3. Run-on and Fused Sentences:** The fused sentence jams together two independent clauses with no punctuation at all. A series of fused sentences or a combination of fused sentences and correctly punctuated sentences creates a run-on sentence.

Fused Sentence Fault: We canceled the picnic it rained that day and my car broke down.

Correction: We canceled the picnic because it rained that day. Then my car broke down.

In contrast to a fused sentence, which can be identified by the absence of punctuation, a run-on sentence cannot be defined precisely. In general terms, a run-on goes on TOO LONG. But how long is too long? Use your best judgement. Consider the context. There is no absolute rule which defines *too long.*

To correct a run-on sentence, break it into smaller sentences.

See *clauses.*

**4. Agreement of Pronouns:** Pronouns agree with their antecedents in person, number, and gender. *To agree* means to correspond in form.

Fault: Jones, Newton, and Perez noted positive correlation of results; in particular, *his* studies showed. . . [*whose?*] (Plural antecedent, singular pronoun)

Correction: Jones, Newton, and Perez noted positive correlation with their results; in particular, Newton's studies showed . . .

See *Pronouns.*

**5. Agreement of Verbs:** A verb agrees with its subject in person and in number, no matter what the subject is and wherever the subject is located in the sentence.

Fault: Many varieties of plankton lives in the sea.
Correction: Many varieties of plankton live in the sea.
[*Varieties,* not *plankton,* is the subject of the verb *live.*]

See *Verbs.*

**6. Dangling Modifiers**: Dangling modifiers are verbal clusters or elliptical adverbial clauses which are incorrectly positioned apart from what they modify, or which possess nothing in the sentence to modify. Dangling modifiers are especially prevalent in sentences beginning with -*ing* words (participles) which do not refer to the subject of the main clause of the sentence.

Rule: The phrase must modify the word it is attached to.

Fault: *Having received your proposal this morning*, **it** is expected that the Director will have time to review it later today.

Correction: *Having received your proposal this morning*, **we** expect that the Director will have time to review it later today.

See *Subordination*.

# Grammar and Mechanics

In the following section, the elements of grammar will be outlined, starting with the most basic concepts. If you already understand what a complete sentence or a subordinate clause is, turn directly to the heading which pertains to your problem. If you need more briefing on general language terms, continue reading.

There are two ways to classify words: by identity and by function.

*Id*entity, *id*entification, *id*entify, *id*—all these words come from the same root.

*Identity:* the collective aspect of the set of characteristics by which a thing is definitively recognized or known.

Words have an identity—what they are. You might remember from grade school that you learned the parts of speech. The parts of speech—*noun, pronoun, verb, adverb, adjective, preposition, conjunction*—name the identities of words.

However, though words possess fixed identities, they can function in different ways in a sentence. Function refers to the way a word behaves in a sentence: what role it takes, what semantic position it occupies. *Subject, predicate,* and *object* name some of the functions a word can have.

Sometimes the identity and the function are the same. An adjective is always an adjective. No matter how you uses the word *red*, it must always function as an attribute describing a substance: a *red* table, a *red* sky, *red* blood. It is born an adjective, will die an adjective, and may in its lifetime hope to be part of a compound adjective—*red*-blooded person—and to describe many nouns.

Sometimes the identity and the function differ. A verb can sometimes behave as a noun. Take *swimming*. In the sentence, "Swimming is my favorite sport," *swimming* functions as a noun in the subject of the sentence.

In the sentence, "He is swimming across the pool," *swimming* is a verb. The word has an identity but its function can change.

For practical purposes, we will divide the identity of the word—*what it is*—from its function—what it does.

| *What It Is* | *What It Does* |
| --- | --- |
| Noun/Pronoun | Subject, Direct Object, Indirect Object, Object of Preposition |
| Verb | Predicate |
| Adjective | Adjective |
| Adverb | Adverb |
| Preposition | Preposition |
| Conjunction | Conjunction |
| Interjection | Interjection |

## NOUN

A *noun* is a word that names a person, place, thing, or abstraction. An abstraction can include an emotion, idea, concept, wish, passion, or quality.

*Person*: scientist, golfer, poet, gossip, rogue, zealot.

*Place*: house, laboratory, cellar, attic, patio, rotunda.

*Thing*: pipet, cigar, wrench, paradox, nosebleed, ice.

*Abstractions*: realism, truth, attraction, beauty, ethos.

There are two sub-classes of nouns: proper nouns and common nouns. *Proper nouns* name a particular person, place, thing or idea: Oedipus, Marie Curie, Socrates, Columbia, Gabriel Garcia Marquez. Proper nouns are always capitalized. A *common noun* names a thing in general: microscope, accordion, moose, nucleus.

A *collective noun*, a subset of common nouns, names a group of things: orchestra, gang, troop, galaxy, squad, cluster, coven, bevy, tribe, audience.

**Function**

A noun, like a pronoun, may function as the subject of a sentence, the direct object, the indirect object, or the object of a preposition.

Subject: The <u>chair</u> is in the room.

Direct Object: He put the <u>chair</u> in the living room.

Indirect Object: He gave (to) the <u>chair</u> a new paint job.

Object of Preposition: He put the book on the <u>chair</u>.

## PRONOUN

A pronoun takes the place of a noun. It must agree with the noun in person, number, and gender.

I (first person, singular) me, mine, my

we (first person, plural) us, ours, our

you (second person, singular) you, yours, your

you (second person, plural) you, yours, your
he/she (third person, singular) him, her, it; his, hers, its; his her, it
they (third person, plural) them, theirs, their
who, whom, whose, whose

The pronoun is inflected in its use.

I have a book. subject
It belongs to me. object
It is mine. predicate adjective
It is my book. possessive

**More Complicated Cases.** The first instance applies to subject-verb agreement as well.

(1) **Agreement with Groups.** Some nouns can take either a singular or plural verb and pronoun. For instance, when the noun is a collective noun, it may act either as an individual or as a group. Such words, which are single words, refer to a group. A word like *colony* refers to one unit composed of many individual parts. Other words like this include the following: *team, colony, United States, pair, scissors, group, class, committee, company, board* [of directors], etc.

To determine agreement, look at the meaning of the sentence: If the group functions as one unit, it is singular. If the sentence emphasizes the individual members of the group, it is plural.

The baseball *team* that participated in the parade were the California Angels; *they* won the game.

The baseball *team* was the California Angels; *it* was coached by Alexander the Great.

(2) **The Universal Subject.** Traditionally the rule in English has been to use the third person masculine pronoun as the universal subject in cases when you do not know the gender or in cases when the gender is mixed. For example, this rule would say, "Everyone in the class has *his* book," even if the group were mixed in gender. Because the masculine pronoun is no longer universally accepted as the common gender, people very frequently try to avoid this phrasing. The most common correction of this problem has been to phrase the sentence as "Everyone in the class has *their* book." Unfortunately, this correction substitutes one error for another because the subject of the sentence, *everyone*, is singular [every *one*]; it must agree with the pronoun in number but *their* is plural. This phrasing would, however, be correct if the sentence referred to a situation in which the class owns one book and *their* book shows collective ownership: one book owned by a class. You have four options in cases such as these: (1) rephrase the sentence entirely: *The class brought their books*; (2) use both pronouns: *Everyone has his and her book*; (3) use the traditional universal subject; (4) make the feminine the universal subject.

## VERBS

A verb is a word which describes an action or a state of being.

**"Be" Verbs:** The follow verbs denote a state of being: *am, is, are, was, were, appear, tastes, seems.*

**Transitive Verbs:** Some verbs convey action. The action passes through the verb to an object. These are called TRANSitive verbs because the action is TRANSmitted THROUGH the verb into the object. In the example

Joe kicked the ball.

the action of kicking starts with Joe, passes through kick and lands on the ball.

**Intransitive Verbs:** Some verbs convey action but cannot be forced under any conditions to relate to an object.

Example: I sit down. I stand up.

Such verbs of self-propelled action do not carry an action through a verb. They have no object. *Lie* ( as in I lie down) is intransitive. It is conjugated as *lie lay lain*. In contrast *lay* is a transitive verb, because it is possible to lay a object down, as in, "I lay my book on the table." Lay is conjugated *lay, laid, laid*. (The verb "to lie" as in to place differs from "to lie" as in to tell a falsehood. This form of lie is conjugated as *lie, lied, lied*.)

### Tense

All verbs have two attributes: tense and voice. Tense denotes time.

| | |
|---|---|
| Present | Joe kicks the ball |
| Present continuous | Joe is kicking the ball. |
| Past | Joe kicked the ball. |
| Present Perfect | Joe has kicked the ball. |
| Present Perfect Continuous | Joe has been kicking the ball. |
| Past Perfect | Joe had kicked the ball. |
| Past Perfect Continuous | Joe had been kicking the ball. |
| Future | Joe will kick the ball. |
| Future Perfect | Joe will have kicked the ball. |
| Future Continuous | Joe will be kicking the ball. |
| Future Perfect Continuous | Joe will have been kicking the ball. |

### Voice

Transitive verbs also possess voice. There are two voices, active and passive. In the active voice, the subject transmits an action through the verb to the object; in the passive voice, the subject receives the action transmitted through the verb.

Compare the active to the passive:

Active: Joe kicked the ball.

Passive: The ball was kicked by Joe.

Notice that the passive takes more words than the active voice to say the same thing. The active voice requires *four* words; the passive voice requires *six* words. Sometimes the passive is said to be wordy because it takes more words in the sentence.

In English, it is appropriate to delete the agent of the passive voice.

Passive with agent deleted: The ball was kicked (by Joe).

The sample was measured (by me).

Because the agent is implied but deleted, in some cases the passive voice may be vague. This means that the audience cannot tell what or who the agent is. The passive voice is appropriate, even desirable, for use in lab reports which describe a process in which the human agent is not the focus of the writing even though it reports what the agent did.

## VERBALS

A *verbal* is a word which is made from a verb but which behaves like a noun or an adjective. There are three kinds of verbals: participials, gerunds, and infinitives.

*Participials.* A Participial is made from a verb but acts like an adjective. It is made from the *participle* form of the verb. (Notice the difference between the two words: participle and participial. Participle is the name of the verb form; participial is the term given to what you make of that verb form.) Verb participles come in two forms, the past and the present.

(1) Past participle of the verb (-ed for regular verbs): received, completed.

Example: I have completed my homework.

(2) Present participle of the verb (-ing form): freezing, smoking.

Example: Paul is smoking ham for Easter.

When you take the participle form of the verb and transform it into a adjective, you have a participial.

Put your completed essays on my desk. [The participial *completed* modifies *essays*.]

Contrast with the *verb* form: I *completed* the essays.

The received truth about this issue is written here. [The participial *received* modifies *truth*.]

Verb: I *received* a telegram.

The freezing rain made the roads dangerous. [The participial *freezing* modifies *rain*.]

Verb: I *froze* lemonade in the freezer.

She left the smoking gun behind. [The participial *smoking* modifies *gun*.]

Verb: He *smoked* three cigarettes.

**Gerunds.** A gerund is word derived from the participle of the verb but it functions as a *noun*.

Example: Ann loves <u>swimming</u>.

*Swimming* is derived from the present participle of the verb "to swim" as in the example, "She is swimming across the pool." In the example, *swimming* behaves as a noun and, in the sentence, it occupies the place of the direct object.

<u>Studying</u> always bored Michael.

*Studying* is derived from the present participle of the verb "to study" as in the example, "Michael is studying in the library." In the example, *studying* behaves as a noun, and in the sentence, it takes the position of the subject.

**Infinitives**: An infinitive is the "to" form of the verb. The infinitive may function as noun, adjective, or adverb.

He plans <u>to go</u>.
He is the man <u>to elect</u>.
Some problems are difficult <u>to solve</u>.

NOTE: You may have encountered this grammatical rule about the infinitive: *Never split an infinitive.* A split infinitive is one in which another word or phrase separates the "to" from the verb.

Example: I wanted <u>to</u> quickly <u>finish</u> my homework.
Example: He wanted <u>to</u> not <u>eat</u>.

*Explanation: quickly* and *not* split up the infinitive because they come between "to" and the verb.

While the rule prohibiting the split infinitive is fading into disuse, some people view split infinitives as a social indecorum. The rule forbidding a split infinitive comes from the time when Latin was the universal language of the world. All scholarly, respectable writing was done in Latin. Scientists and scholars even took Latin names to show that they were learned. In Latin, infinitives appear as a single word. The rule which prohibits splitting an infinite shows deference to Latin and to the time when the rules which governed Latin grammar were applied to other languages.

For your purposes, you should decide on your own whether or not you will split an infinitive. On the one hand, it comes from an archaic and perhaps even absurd reasoning; on the other hand, some people judge the "proper" use of the infinitive as a mark of a person who is experienced in language use. The decision is yours.

## PREPOSITIONS

A preposition is a word which denotes a position in space or time. An easy rule of thumb is this: a preposition is anything a plane can do to a cloud or anything that a sheep can do with a fence. Prepositions: among, around, before, beneath, between, in, on, on top of, through, under.

## CONJUNCTIONS

Conjunctions function like hinges. They are words which link other words together. Conjunctions include *and, or, but, nor, for, so,* and *yet.*

## ADVERBS

Adverbs are words that modify

—a verb. He ran <u>quickly</u>.
— an adjective. The surprise made me <u>very</u> happy.
— another adverb. He ran <u>very</u> quickly.

# From Words to Sentences

## Clauses

*Definition*

A clause is a unit of meaning which contains a subject and a predicate.
The predicate is a verb; the subject may be a noun or a pronoun.
No object is required to make a clause, but a clause could have one.
There are two kinds of clauses:

(1) main (independent) clause
An independent clause contains a complete thought.
Example: I will leave.
(2) subordinate (dependent) clause

A dependent clause must be attached to a main clause to complete its meaning.

Example: When the time is right. . . .(what will happen?)

*Clauses as Sentences*

A sentence must express a complete thought and must contain a subject and a verb. "Close the door" is a sentence because, in the imperative, the subject "you" is implied.

*Grammatical Error:* A piece of a sentence does not express a complete thought. Pieces of sentences are called *fragments*. Do not write sentence fragments. One reason students write fragments is because they confuse a verbal with a verb.

*Fragments:* Being that it was time to go. The red painted pony.

*Subject-Verb Agreement*

1. The subject must agree with the verb in person and number. English

is an irregularly inflected language; the most common variation in the verb inflection is in the third person singular.

| Person | singular | plural |
|--------|----------|--------|
| first | I go | we go |
| second | you go | you go |
| third | he, she, or it goes | they go |

2. Do not be misled by words or phrases that intervene between the subject and the verb. In your mind, cross out prepositional phrases which modify the subject to isolate it.

Incorrect:    The <u>cars</u> in the lot <u>looks</u> broken-down.
Correct:      The <u>cars</u> in the lot <u>look</u> broken-down.

3. Usually subjects joined by *and* are plural.
Example:       <u>Oxygen and hydrogen</u> are elements.
*Exception:* When *and* is used to join words in a unit, the unit is singular.

Example: <u>Bacon and eggs</u> is my favorite breakfast.

4. *Each* and *every* are singular: These words imply the meaning each (one) and every (one). Even when the subject is compound, with elements joined by and, the verb is singular:

Example: <u>Each nut</u> and <u>every bolt is</u> tested for accuracy.
5. Singular subjects joined by *either/or* or *neither/nor* take a singular verb.

Example: <u>Neither</u> José <u>nor</u> Maria <u>is</u> coming to the party.
But if one subject is singular and one is plural, the verb agrees with the subject which is closest.

Example: Neither the substance nor the <u>conditions are</u> changed.

6. Do not be misled by inverted word order (verb + subject) or (there + verb + subject).

Examples: There <u>was time</u> to accomplish everything.
There <u>were winds</u> from the east blowing through the hills.

7. A linking verb agrees with its subject, not with its complement.

Example: His <u>problem is</u> frequent headaches.
Example: <u>Frequent headaches are</u> his problem.

8. Nouns plural in form but singular in meaning take a singular verb.

Example: <u>News travels</u> fast.

Frequently used words in this category are *news, mumps, measles, economics, electronics, physics.*

*Kinds of Clauses*

Clauses can be arranged into four kinds of sentences.

1. *Simple Sentence* = one independent clause
   Typical paradigm: subject + verb + direct object
   The protozoan parasite causes Chagas' disease.

2. *Compound Sentence* = two or more independent clauses connected by a coordinating conjunction

(The protozoan parasite causes Chagas' disease)
and
(it creates a health hazard affecting 20 million people in Central and South America).

3. *Complex Sentence* = one independent clause + one or more dependent clauses connected by a coordinating conjunction

| | |
|---|---|
| independent | (The failure to control the spread of the infection has been attributed to the parasite's life cycle) |
| dependent | (which has different morphological forms in both the insect vector and the vertebrate host.) |

4. *Compound-Complex Sentences* = at least two independent clauses (the compound part) + one or more dependent clauses (the complex part)

| | |
|---|---|
| independent | (Upon transmission to mammalian host, the infective parasite circulates in the bloodstream as a nondividing form of parasite,) |
| dependent | (until infection is established by invasion in host cells); |
| adverb | (then) |
| independent | (a subsequent transformation to host the intracellular amastigote occurs.) |

*Use of Clauses*

1. The audience needs will dictate the kind of sentences you make.
2. Vary the kind of sentences you use.
3. Avoid packing many dependent clauses into one sentence.

*Significance*

Sentence structure shows a relationship between ideas which are equal (compound) or hierarchical (complex and its forms). Complex sentences show a relationship of subordination; one concept is more important than another. The main idea takes the independent clause; the subordinate concept takes the dependent clause.

Coordinate means "being of equal structural rank."
Subordinate means "being of lower structural rank."

## Coordination

1. Use coordination to give equal ideas equal emphasis.
(The offer was tempting) but (I didn't accept it.)
2. A compound sentence shows a stronger connection between ideas than does two separate sentences.
3. There are two ways to connect independent clauses.
  A. add a comma + coordinating conjunction (*and, or, but, nor, for, so, yet*)
  B. use a semicolon between independent clauses
     A semicolon always connects independent clauses with one exception.
     Exception: Use a semicolon to separate a series of items which contain commas.
     Example: The three scientists at the meeting were Dr. Eric Han, Kennedy Space Center; Dr. Hermes Trismigestus, Library of Alexandria; and Dr. Sheetal Kundalini, former director of the Rejkavi Institute of Applied Grammatology.

*Grammatical Errors:*
1. If you link two independent clauses with only a comma, you make a *comma splice.*
2. If you link two independent clauses together with no punctuation, you make a *fused sentence.*
3. Do not string main clauses together when some ideas should be subordinated.

Awkward: I wanted to go to college, so I mowed and trimmed lawns all summer, and that way I could earn my tuition.
Better: Because I wanted to go to college, I mowed and trimmed lawns to earn my tuition.

## Subordination

1. Subordination signals to the reader that one point (expressed in the main clause) is more important than another (expressed in the dependent clause).
The protozoan parasite, which causes Chagas' disease, creates a major health hazard.

2. A subordinate clauses always possesses a specific relationship to the main clauses. A subordinate clause specifies time, place, cause, concession, condition, exception, purpose or description. In the examples below, brackets surround the dependent clause.

| | |
|---|---|
| *Time* | Everyone panicked [when the earthquake struck]. |
| *Place* | Greenwich Time was set in London (where the Royal Society was founded). |
| *Cause* | [Because the tissue damage caused by free radicals is not limited to DNA,] defense mechanisms against free radicals will always be a subject of interest. |
| *Concession* | [Although past studies have shown this to be true,] our results suggest a more tentative approach is merited. |
| *Condition* | We cannot conduct this seminar [unless the university provides us with more computers]. |
| *Exception* | [Except for innovative studies conducted at the Basel Institute for Applied Research,] no reports to date have shown this correlation. |
| *Purpose* | [In order to demonstrate the efficacy of new technology,] our collaborative group devised a new system of graphics. |
| *Description* | Professor Luis Villarreal, [who is famous for his work on pattern recognition,] made up this sentence. |

3. Faulty subordination shows an error in reasoning. Avoid faulty subordination.

Put your main idea into the main clause.

Awkward: Jesus, who was hit by a car, wore a red hat.

Which is more important—*Jesus wore a red hat* or *Jesus was hit by a car*?

Put the important point in the main clause position.

Better: Jesus, who wore a red hat, was hit by a car.

Note: Use context to make a decision about what points are important in the sentence.

# Style

When you revise your writing for stylistic purposes, you polish it up in order to make your point as clearly and concisely as possible. Stylistic revision should be done when you have substantially completed your writing task and want to make it better.

1. Revise according to the need of your audience and purpose.

2. Avoid wordiness. *Wordy* means that you could accomplish the same meaning with fewer words. Some words or phrases are place holders like *zero*; they fill a place in a sentence but carry no semantic content. To avoid wordiness, reduce redundancy. *Redundant* means you do not need a word because you have expressed the point sufficiently already. In other words, you do not have to say the same thing twice. Example: In the phrase, *completely eliminate, eliminate* already contains the meaning of *complete*. Words like *connect, collaborate, cooperate,* or *couple* already imply a relationship between parts, so it would be redundant to say collaborate together. Other redundant expressions (in the brackets) include the following:

yellow [in color]             at 9:45 P.M. [that night]
[basic] essentials           circular [in shape]
return [back]                 small [-size] potatoes
[true] facts

3. Be concise. Do not use many words when a few will express the idea well.

Wordy: In the event that the grading system is changed, expect complaints on the part of the students.
Concise: If the grading system is changed, expect complaints on the part of the students.

Wordy: He spoke in a low and hard-to-hear voice.
Concise: He mumbled.

4. Make verbs do the work rather than nouns. "Sarah <u>has communication</u> with God" versus "Sarah <u>communicates</u> with God."

5. Use action verbs rather than "be" verbs.
Avoid using ("be" verb + adjective) when you *could* use an action verb instead: "There <u>was a meeting</u>" versus "They <u>met</u>."

Note: Sometimes you need to use "be" verbs to express states of being.
(is operative) can be expressed as *operate*
(is beneficial)                *benefits*
(are similar)                  *resemble*
(make contact with)           *contact*

6. Place the most important idea in the most important syntactic positions (subject verb object).

"The departure of the fleet is thought to be necessarily conditional on the weather" versus "Bad weather may keep the fleet at anchor."

7. Avoid delaying phrases/non-referential pronouns. Some phrases are place holders; they usually delay the meaning of the sentence: *It is evident*; *It is obvious that*; *There is, are*.

8. Eliminate pointless elaborate prepositional phrases or connectives:
Wordy: This problem is <u>in the nature of</u> one encountered years ago.
Concise: This problem is <u>like</u> one encountered years ago.
Other wordy or redundant phrases include the following (the most concise version is inside the parenthesis):

     absolutely essential (essential)
     actual experience (experience)
     at the present time (at present, now)
     at this point in time (now, currently)
     completely eliminated (eliminated)
     collaborate together (collaborate)
     during the time that (while)
     few in number (few)
     in many cases (often)
     in most cases (usually)
     in this case (here)
     in all cases (always)
     involve the necessity of (necessitates, requires)
     in connection with (about)
     in the event of (if)
     in the neighborhood of (about)
     make application to (apply)
     make contact with (contact, see, meet)
     maintain cost control (control costs)
     on the part of (by)
     past history (history)
     range all the way from (range)
     red in color (red)
     stunted in growth (stunted)
     subsequent to (after)
     through the use of (by, with)
     true facts (facts)
     until such time as (until)
     with the object of (to)

# Index